Antibody Expression and Engineering

ACS SYMPOSIUM SERIES **604**

Antibody Expression and Engineering

Henry Y. Wang, EDITOR
University of Michigan

Tadayuki Imanaka, EDITOR
Osaka University

Developed from a symposium sponsored
by the Division of Biochemical Technology
at the 207th National Meeting
of the American Chemical Society,
San Diego, California,
March 13–17, 1994

American Chemical Society, Washington, DC 1995

Library of Congress Cataloging-in-Publication Data

Antibody expression and engineering / Henry Y. Wang, editor, Tadayuki Imanaka, editor.

p. cm.—(ACS symposium series, ISSN 0097–6156; 604)

"Developed from a symposium sponsored by the Division of Biochemical Technology at the 207th National Meeting of the American Chemical Society, San Diego, California, March 13–17, 1994."

Includes bibliographical references and indexes.

ISBN 0–8412–3314–4

1. Monoclonal antibodies—Biotechnology—Congresses. 2. Immunogenetics—Congresses. 3. Immunoglobulin genes—Congresses. 4. Hybridomas—Congresses. 5. Gene expression—Congresses.

I. Wang, Henry Y., 1951– . II. Imanaka, Tadayuki, 1945– . III. American Chemical Society. Division of Biochemical Technology. IV. American Chemical Society. Meeting (207th: 1994: San Diego, Calif.) V. Series.

[DNLM: 1. Antibodies—genetics—congresses. 2. Gene Expression—congresses. 3. Genetic Expression—congresses. 4. Genetic Engineering—methods—congresses. QW 575 A629 1995]

TP248.65.M65A56 1995
616.07′88—dc20
DNLM/DLC
for Library of Congress

95–33161
CIP

This book is printed on acid-free, recycled paper.

1995 Advisory Board

ACS Symposium Series

Foreword

THE ACS SYMPOSIUM SERIES was first published in 1974 to provide a mechanism for publishing symposia quickly in book form. The purpose of this series is to publish comprehensive books developed from symposia, which are usually "snapshots in time" of the current research being done on a topic, plus some review material on the topic. For this reason, it is necessary that the papers be published as quickly as possible.

Before a symposium-based book is put under contract, the proposed table of contents is reviewed for appropriateness to the topic and for comprehensiveness of the collection. Some papers are excluded at this point, and others are added to round out the scope of the volume. In addition, a draft of each paper is peer-reviewed prior to final acceptance or rejection. This anonymous review process is supervised by the organizer(s) of the symposium, who become the editor(s) of the book. The authors then revise their papers according to the recommendations of both the reviewers and the editors, prepare camera-ready copy, and submit the final papers to the editors, who check that all necessary revisions have been made.

As a rule, only original research papers and original review papers are included in the volumes. Verbatim reproductions of previously published papers are not accepted.

Contents

INDEXES

Preface

ANTIBODIES ARE MULTIDOMAIN GLYCOPROTEIN molecules of intense interest to the biotechnology community. Antibodies are immunoglobulin molecules of the immune system developed by higher organisms to combat the invasion of foreign substances (antigens). Each antibody has specific activity for the foreign material that elicits its synthesis. In nature, each antibody is made by a different line of lymphocytes and their derived plasma cells in a specific animal. If one could pick out such a cell making a single specific antibody and grow it in a culture, then the cell's progeny, or clone, would be a source of large amounts of identical antibody against a single antigenic determinant: a monoclonal antibody (MAb*). Unfortunately, antibody-secreting cells cannot be maintained well in a culture medium.

In 1975, Kohler and Milstein fused mouse myeloma cells with lymphocytes from the spleen of mice immunized with a particular antigen. The resulting hybrid myeloma, or "hybridoma," cells expressed both the lymphocytes' property of specific-antibody production and the immortal character of myeloma cells. Since then the hybridoma technology has been widely adopted as a method of choice for the preparation of MAbs to a wide spectrum of antigens. Various hybridomas can be cloned and grown in large quantity for indefinite periods of time, and they secrete high concentrations of monoclonal antibodies. Depending on the source of lymphocyte and myeloma cells and by using additional powerful genetic manipulation techniques, different combinations of hybrids (mouse–mouse, mouse–human, and human–human hybridomas) have now been synthesized.

Some of the real or potential applications of monoclonal antibodies will require kilogram quantities of highly purified antibody. Therefore, cost-efficient methods for producing large quantities of highly purified monoclonal antibodies must be developed. Process development will be an important factor for the successful commercialization of any of the therapeutic monoclonal antibodies. Through better understanding of the nutritional requirements of these cells, many improvements in culture media and feeding strategies have been developed. Costly and undefined serum components have been eliminated from media formulation without

*A number of abbreviations for monoclonal antibodies are used in this book: MAb, Mab, Abs, MoAbs. They all denote the same thing.

impairing the immunoglobulin secretion or the consistency of the antibody properties. Development of serum-free media has also made the antibody-manufacturing process more economically competitive.

Many antibodies have the potential to be the most economically significant products derived from mammalian cells. They are already being used extensively in diagnostic assays and their therapeutic potential is increasing with the development of affinity purification systems and in vivo imaging. The immunotherapeutic application uses monoclonal antibodies either as so-called "magic bullets" to destroy malignant tissue (cancer therapy) or as target-specific drug delivery agents (for immunological diseases).

Recent progress in our ability to genetically manipulate and alter the antibody molecules and then express the whole antibody molecule or its selected fragments in various recombinant hosts has profoundly altered the fields of protein chemistry and molecular immunology.

About the Book

In the first few chapters of this book, recent advances in antibody expression and production of several industrially important monoclonal antibodies based on traditional myeloma cells are described. Using a case-study approach, David Robinson and his colleagues show how gene amplification, medium optimization, and process development can be used to optimize a humanized anti-CD18 antibody. Almost 1 g/L of monoclonal antibody can be routinely produced in an optimized fed-batch culture. Sawada and Kitano describe the development and characterization of a human monoclonal antibody against hepatitis B virus surface antigen (HBsAg). Humanization of the MAbs is now routinely carried out to design therapeutics that minimize the formation of human anti-mouse antibodies (HAMA) for various in vivo applications. Besides using myeloma cells as hosts, M. E. Reff's group has developed an efficient expression system to produce whole immunoglobulin molecules in Chinese hamster ovary (CHO) cells. Almost 800 mg/L of antibody was achieved in a 100-L fermentor in 6 days. This is quite competitive with the traditional myeloma hosts.

Genetic shuffling of antibody domains and other active molecules can also be used to create new chimeric molecules with various combinations of binding and effector functions and novel antigen-binding molecules. Similarly, recombinant DNA techniques have now been used to produce different antibody fragments with reduced molecular mass, such as Fab, Fv, or single-chain antigen-binding molecules. These engineered monoclonal antibodies and their corresponding fragments have the potential to expand on various clinical applications, such as diagnostic imaging and immunotherapeutics, that require smaller molecules for deeper penetration or with altered binding properties. The next few chapters are devoted to the expression of monoclonal antibody and its fragments in other non-mammalian host systems. Potter et al. demonstrate the use of baculovirus expression system to produce various forms of hybrid or chimeric antibodies of IgG, IgM, and IgA isotypes in insect cells. Julian Ma reports on the progress of expressing antibody fragments in cultivated plant cells and plants.

Microbial gene expression systems have also been used routinely to produce antibody fragments (such as Fab and Fv) consisting of one or two antibody chains. Co-expressing antibody light and heavy chains in various microbial hosts have been demonstrated since 1984. David Filpula and his colleagues review the potential of using the *Escherichia coli* expression system to overproduce a single-chain Fv protein with human IgM $C\mu$ specificity. Better and Nolan describe the construction of a family of immunofusion proteins linking various antibody fragments of the T-cell targeted H65 antibody with the plant ribosome-inactivating protein called gelonin in *E. coli*. Some of these fusion proteins are as potent as some corresponding chemical conjugates developed for use as immunotoxins. S. L. Wong and his group have shown that a multiple-protease-deficient Gram-positive *Bacillus subtilis* can be used as an alternate expression host to produce an anti-digoxin single-chain antibody fragment. Nyyssonen and Keranen discuss the use of cellulase-producing *Trichoderma reesei* as a promising host to produce a Fab fragment fused to the cellulase gene.

In the remaining chapters, Lavey and Janda review the exciting and promising field of catalytic antibodies and their applications in organic synthesis. For a more specific application, T. Imanaka's group have successfully generated a light chain of a MAb against porphyrin that still retains peroxidase-like activity.

This book covers many recent developments in antibody expression and engineering using traditional hybridoma and nontraditional recombinant host systems and a few potential applications. It is not intended to be a comprehensive handbook. Increasingly, fine details of the antibody molecule have become known through X-ray crystallography and NMR spectroscopy. Thus, quantitative relationships between structure and function can now be estimated and manipulated through genetic engineering. Similarly, screening for novel antibodies through the generation of combinatorial libraries using filamentous phage and other microbial antibody expression systems is becoming routine. Therefore, the advances in genetic manipulation and process development discussed in this book coupled with other advances will propel the field of antibody engineering to a new horizon.

Acknowledgments

We gratefully acknowledge the assistance of many anonymous reviewers who helped us carry out in-depth reviews and supplied important and constructive criticism of the original manuscripts. We also acknowledge the patience and support from the staff of the ACS Books Department, particularly Barbara Pralle, that made this book a reality.

HENRY Y. WANG
Department of Chemical Engineering
University of Michigan
Ann Arbor, MI 48109–2136

TADAYUKI IMANAKA
Department of Biotechnology
Faculty of Engineering
Osaka University
Yamadaoka, Suita
Osaka 565, Japan

June 14, 1995

Chapter 1

Production of Engineered Antibodies in Myeloma and Hybridoma Cells

Enhancements in Gene Expression and Process Design

D. K. Robinson, D. DiStefano, S. L. Gould, G. Cuca, T. C. Seamans, D. Benincasa, S. Munshi, C. P. Chan, J. Stafford-Hollis, G. F. Hollis[1], D. Jain[2], K. Ramasubramanyan, G. E. Mark, and M. Siberklang[3]

Department of Cellular and Molecular Biology, Merck Research Laboratories, Rahway, NJ 07065

As reviewed in this report and exemplified in a case study, recombinant antibodies (Abs) can be expressed to very high levels in lymphoid cells by optimizing gene expression and process conditions. Myeloma and hybridoma cell lines have been constructed that secrete functionally active, recombinant antibodies at up to 15 pg/c/d in non-amplified cell lines and up to 100 pg/c/d in gene-amplified cell lines. Processes have been developed that achieve final titers in fed-batch culture of 1-2 g/L, or reach volumetric productivities in perfusion culture of 0.2-0.6 g/L/d. In the specific example of the production of a humanized anti-CD18 Ab, h1B4, amplified NS0 cell lines were developed that secreted Ab at up to 50 pg/c/d. Periodic addition of concentrated nutrient solutions in fed-batch culture allowed production of the Ab to 1.8 g/L. In perfusion culture, titers of up to 0.5 g/L were obtained with maximum volumetric (*i.e.* reactor volume) productivities in excess of 0.6 g/L/d.

Production of Engineered Antibodies

Because of their extraordinary molecular recognition capabilities and exquisite specificity, Abs have been considered for many potential therapeutic interventions, including (as discussed below) the short-circuiting of the inflammatory reaction. Abs can be engineered via recombinant DNA technology to optimize their therapeutic efficacy or expression levels (as reviewed in *1-2*). For example, the non-binding domains of non-human Abs can be replaced by the corresponding human sequences to minimize potential human antimouse antibody (HAMA) responses (*3-4*); the isotype of the constant region can be changed to modulate effector function interactions (*4-5*); the binding domains can be modified to enhance affinity or specificity (*6*); or Abs of the appropriate specificity can be selected from cloned *E. Coli* or phage combinatorial libraries (*7-8*). Such Abs can be efficiently expressed with proper post-translational modification in a number of mammalian cell systems, the most common being Chinese Hamster Ovary cells (*9-11*) and, as

[1]Current address: DuPont Merck Pharmaceutical Co., Glenolden, PA 19036
[2]Current address: R. W. Johnson Pharmaceutical Research Institute, Raritan, NJ 08869
[3]Current address: Enzon Inc., 20 Kingsbridge Road, Piscataway, NJ 08854

0097–6156/95/0604–0001$12.00/0

reviewed in this manuscript, various cell lines of myeloma or hybridoma origin that have lost expression of their endogenous immunoglobulin (Ig) chains.

Gene Expression. Recombinant Ig genes have been expressed in a number of rodent cell lines (see Table I). In particular, the murine hybridoma line SP2/0 (*12*), and myeloma line NS0 (*13*) have been used to produce Ab from both cDNA and genomic DNA sequences. Transcriptional regulation in most expression vectors has been based on either Ig or viral promoters, such as the Cytomegalovirus immediate early (CMVIE) promoter. In many cases, the light and heavy chain genes are located on separate vectors which are introduced into the host cell genome by either sequential transfection or co-transfection. Since accumulation of unassembled heavy chains has been considered toxic to the cell (*14*), many sequential transfection protocols start with expression of the light chain gene. However, Dorai and Moore have shown that cell lines can be constructed that accumulate up to 2% of their total cellular protein as unassembled heavy chain (*15*), indicating that heavy chains may not be toxic. Larger "tandem" vectors that contain a selectable marker gene and both the heavy and light chain genes in a single plasmid have been transfected into YB2/0 and NS0 cells with high efficiency, generating high expressing cell lines (Table Ia).

In most cases where expression has been directly compared between SP2/0 and other cell lines, including P3.653 (*21*), P3U1 (*17*), and CHO (*19*), the highest expression levels (up to 100 mg/L in batch culture) were seen in SP2/0. However, when Shitara transfected three cell lines, SP2/0, P3U1 and YB2/0 with a tandem vector encoding both heavy and light chain chimeric cDNA, the highest transfection efficiency was seen in the rat myeloma, YB2/0 (*23*). Bebbington et al. designed vectors based on the glutamine synthetase (GS) gene as a selectable marker, conferring the ability to grow in glutamine-free medium. A GS vector was used to express a chimeric anti-Tag72 tumor marker Ab in NS0 using the CMVIE promoter. Glutamine-independent transfectants secreted Ab at up to 4 pg/c/d (*24*). We have also used GS vectors to express Ab in NS0 cells and have achieved specific productivities of up to 15 pg/c/d, reaching batch titers of up to 100 mg/L (*25*).

Gene amplification can be used to derive cell lines that secrete Ab at even greater rates (see Table Ib). As for CHO (see *26* for review), genes cloned adjacent to the dihydrofolate reductase (DHFR) gene can be amplified by selecting for cells resistant to increasing concentrations of the drug methotrexate (MTX). Dorai and Moore isolated transfected SP2/0 cells in the presence of up to 0.5 mM MTX that had increased secretion of Ab by 25-fold, up to 25 pg/c/d (*15*). Shitara used MTX concentrations of 0.05 to 0.2 mM to select an amplified YB2/0 line that secreted Ab at rates of up to 100 pg/c/d (*23*), equivalent to the highest expression levels reported for amplified CHO cell lines (*10-11*). The substrate analog, methionine sulfoximine (MSX), can be used to amplify GS vectors in NS0. Selection with 20-100 uM MSX resulted in cell lines that expressed two-fold more Ab, 10 versus 4 pg/c/d (*24*). Similarly, we have derived amplified NS0 that express Ab at rates of up to 50 pg/c/d as described below.

Hybridoma cell lines can secrete endogenous Ab very efficiently. Specific productivities range from less than 2 up to 80 pg/c/d (*27*), the cumulative Ab secreted over a 24 hr. period representing more than 50% of the total cellular protein. Similarly, lymphoid cell lines, such as SP2/0 or NS0, can efficiently express transfected recombinant Ig genes (Table I). The genomic integration of transfected vectors is generally a random event (*28*). However, targeting of expression cassettes to the normal chromosomal site for Ig genes can be exploited to efficiently generate high level expression clones. Shulman and coworkers first showed that normal IgM expression could be restored to non-expressing hybridoma cell lines by homologous recombination (*29*), yielding cell lines that secreted Ab at near endogenous levels, in this case 2.6 pg-IgM/c/d (*30*). In our own work, we have identified primary transfectant NS0 lines that secrete recombinant Ab at high

Table I. IgG gene expression in myeloma cell lines

A. Non-amplified cell lines

Cell Line	Transfection Order/DNA	Selectable Marker H	Selectable Marker L	Promotor/ Enhancer	Specific Productivity (maximum) pg/c/d	Final Batch Titer mg/L	Reference
X63Ag8.653	seq/cDNA	Hyg	Neo	CMV	---	0.4	*16*
P3-U1	co/genomic	gpt	Neo	SV40	1	---	*17*
YO	co/genomic	gpt	Neo	SV40	5	40	*18*
SP2/0	co/genomic	Neo	Neo	mIg	36 pg/c/mL*		*19*
SP2/0	co/genomic	Neo	gpt	mIg	4	6	*20*
SP2/0	seq/genomic	gpt	Neo	mIg	32pg/c/mL*	100	*21*
SP2/0	single-chain/cDNA		Neo	CMV	---	4	*22*
SP2/0	co/genomic	gpt	Neo	SV40	3	---	*17*
YB2/0	tandem/cDNA		Neo	MV	---	---	*23*
NS0	tandem/cDNA	gs		CMV-IE	4	---	*24*
NS0	tandem/genomic	gs		CMV-IE	15	100	*25*

B. Amplified cell lines

Cell Line	Transfection Order/DNA	Amplifiable Marker	Selective Agent	Specific Productivity (maximum) pg/c/d	Final Batch Titer mg/L	Reference
SP2/0	co/genomic	DHFR	MTX	25	---	*15*
YB2/0	tandem/cDNA	DHFR	MTX	100	---	*23*
NS0	tandem/cDNA	GS	MSX	10	35	*24*
NS0	tandem/genomic	GS	MSX	50	140	*25*

*Batch titers normalized to cell number (pg/c/mL), not specific productivity.

Abbreviations used -- CMV: cytomegalovirus immediate early promotor; mIg: murine immunoglobulin; MV: Moloney virus long terminal repeat; gpt: xanthine-guanine phosphoribosyl transferase; MTX: methotrexate; MSX: methionine sulfoxamine; Neo: neomycin; Hyg: hygromycin; gs: glutamine synthetase; co: co-transfection of heavy and light chain vectors; seq: sequential transfection with heavy and light chain vectors; tandem: transfection of vector containing both heavy and light chain DNA sequences.

rates, including one where the presence of only a single genomic copy of the expression vector was confirmed by Southern blotting. On the assumption that the site of chromosomal integration permits high level transcription, we isolated the insertion site of the expression vector from this cell line and used it to target integration of expression vectors to the same site in wild type NS0 cells. Application of such targeting vectors improved the frequency of high-expression clones by up to four-fold. These clones can express extremely high levels of recombinant Ab (15-40 pg/c/d) from a single copy of the inserted expression vector (Benincasa, D. et. al., manuscript in preparation).

In summary, the highest expression levels have been reached in either CHO cells or in two murine and one rat lymphoid cell lines, SP2/0, NS0, and YB2/0, respectively. High expression recombinant cell lines were obtained by all three transfection protocols -- sequential, co- and tandem-vector transfection, using either cDNA or genomic Ig genes. The very highest expressing cell lines have all resulted from amplification of tandem vectors.

Process Design. Typically, Ab production in batch cultures (Table I) is limited by the depletion of an essential nutrient (*31*) or generation of metabolic byproducts to inhibitory levels (*32*). Two main approaches have evolved to maximize production in the face of these limitations: 1) the use of fed-batch culture to replenish depleted nutrients while minimizing byproduct formation or, 2) the use of perfusion culture to continuously supply fresh medium while removing spent medium and byproducts (*33*).

High Ab titers in fed-batch culture can be reached by maximizing the total culture productivity, as best measured by the compound effects of increasing the culture viability index (*34*) and the specific productivity of the cells (*35*). In serum-free batch air-lift reactors, Bebbington et al. showed that an amplified NS0 line grew to just over 10^6/mL and secreted Ab up to a final concentration of 350 mg/L; addition of an amino acid feed solution increased the maximum cell density and specific productivity three-fold, to 3 x 10^6/mL and 35 pg/c/d, respectively. Ab accumulated to 560 mg/L (*24*). Maiorella and coworkers demonstrated that monoclonal Ab yields of 1 g/L could be achieved in fed-batch culture of hybridoma cells by optimizing the basal medium and carefully designing the composition of the nutritional feed solution; adding combinations of glucose, amino acids, phospholipid precursors, disulfide exchange reagents, vitamins, and trace elements to maintain viable cell densitities of more than 1x10^6/mL for up to 300 hr. (*36*). As described below, we have added concentrated nutrient solutions to serum-free cultureds of an amplified NS0 cell line, yielding final Ab titers of 1.8 g/L.

Perfusion culture can also be used to supply fresh nutrients to cells (*33*), including transfected hybridoma cells (*37*). In batch culture, an SP2/0 cell line expressing a chimeric (mouse variable, human constant regions) anti-CD7 Ab reached final titers of 100 mg/L with a volumetric productivity (based on the reactor volume) of approximately 12 mg/L/d. When this cell line was grown in continuous (chemostat) culture, the titer was maintained at 100 mg/L and the volumetric productivity was increased nearly ten-fold to 100 mg/L/d. In this case, the culture dilution rate was limited by the maximum growth rate of the cells to 1 volume/d, limiting the maximum achievable volumetric productivity. When a rotating sieve was used to partially retain the viable cells within the reactor (*38*), perfusion rates in excess of the cellular growth rate could be used and the volumetric productivity was further increased to 200 mg/L/d, nearly twenty-fold greater than that achieved in batch culture (*37*). By combining the use of medium perfusion with the addition of concentrated nutrients, both the product titer and the volumetric productivity can be increased even further, as described in the case study below.

Case Study: Production of an Anti-CD18 Humanized Ab in NS0 Cells

The murine monoclonal Ab 1B4 is very specifically directed against human CD-18, a component of the β-2 integrin family that is involved in leukocyte extravasation and migration to sites of inflammation. 1B4 has furthermore been shown to prevent attachment of human neutrophils to human endothelial cells *in vitro* (*39-42*). To explore the application of this Ab as a potential intervention in human inflammatory processes, murine 1B4 was humanized by an efficient series of genetic constructions, whereby the humanized heavy and light variable regions were expressed as fusions to the human IgG4 and Cκ constant regions, respectively (*5, 43-44*). Of the resulting successfully engineered humanized anti-CD18 antibodies, one (h1B4 or Gal/Rei in *5*) was expressed to very high levels in NS0 murine myeloma cells using a GS selectable marker vector (*24*).

Materials and Methods: Humanized 1B4 Antibody Expression in NS0 Cells. The h1B4 heavy and light chain genes (*5, 43-44*) were coexpressed in a single tandem vector, each under CMVIE promoter transcriptional control. The vector also contained the GS selectable marker gene (CellTech, Ltd). NS0 cells were transfected by electroporation and selected in glutamine-free medium. Among the clonal transfectants isolated, D12 was expanded and chosen for gene amplification selection by exposure to MSX (*45*). The highest expressing MSX-resistant clone was termed 1G4. It was expanded and further subjected to single-cell dilution cloning in serum-free medium, resulting in the isolation of F11.

Analysis of Gene Amplification by Southern Blotting. Genomic DNA from clones D12, 1G4, and F11 was isolated by a Proteinase K/sodium-dodecyl-sulfate procedure as recommended by the manufacturer (Applied Biosystems, Foster City, CA). To identify the inserted expression vector, 10 μg of genomic DNA from each of the three clones, as well as the parental cell line, NS0, were digested to completion with XbaI, separated on an 0.8% agarose gel, and transferred to a nitrocellulose membrane by the Southern method (*46*). The blot was hybridized using a 929 b.p. fragment containing the human immunoglobulin kappa constant region gene (*IGKC*) as a probe and bands were visualized by autoradiography. DNA probes for hybridization were labeled by nick translation according to the manufacturer's instructions (GIBCO/BRL, Gaithersberg, MD). Signal intensity was determined on a Phosphorimager (Molecular Dynamics). DNA loading per lane was normalized to the hybridization signal strength of the endogenous mouse immunoglobulin kappa constant region.

Serum-Free Medium Development and Suspension Adaptation. Suspension adaptation of NS0 clones was performed over 2-4 weeks in shake flasks as described (*47*). Because the recombinant clones originated from a suspension-capable parental cell line, this process was usually straightforward and efficient. Concurrently, a serum-free medium formulation was developed by substituting ethanolamine, ß-mercaptoethanol, bovine serum albumin, transferrin, insulin and low density lipoprotein for dialyzed fetal bovine serum in a glutamine-free HEPES-buffered Iscove's basal medium supplemented amino acids and nucleosides (*44*). Suspension/shear adaptation was completed in this serum-free medium containing 0.1% Pluronic F-68, first in shake flasks and finally in spinner flasks. The medium for 1G4 and F11 also contained MSX.

Culture and Scale-Up Conditions. Cells were grown at 37°C in 125 or 250 ml shake flasks (Corning) with a working volume of 10-50 ml, in spinner flasks (Bellco Microcarrier Spinner Flasks) at 100 mL to 25 L working volume, and in a stainless steel stirred reactor (Sulzer/MBR 18-L Spinferm). Shake flasks were agitated at 100 rpm and were either gassed daily with 5% CO_2 and resealed or were maintained in a 5% CO_2 incubator. Small spinner flasks (100-250 mL working volume) were agitated at 100 rpm in a 5% CO_2 incubator. Larger spinner flasks were agitated at 30-80 rpm and dissolved gasses were maintained by continuous headspace gassing of 5% CO_2. The stirred reactor was maintained at 16 L working volume and agitated at 80-90 rpm. The culture pH was maintained by adjusting the headspace gas composition (air, CO_2). The dissolved oxygen levels were maintained by microsparging O_2. Cell retention perfusion was achieved by medium withdrawal through a top-mounted spin filter agitated independently at >110 rpm.

Analytical. Cells were counted microscopically by hemacytometer and viability was determined by trypan blue dye exclusion. h1B4 Ab was measured by ELISA or by Protein A (*48*) affinity chromatography (Oros Protein A and Perseptive Biosystems, Inc. Poros Protein A resin). Amino acid analysis of medium samples was by the PicoTag™ (Millipore, Inc., Bedford MA) pre-column derivatization HPLC method. Glucose, ammonia and lactate were analyzed using a Kodak Biolyzer.

Results: Gene Amplification and Antibody Expression Level. Clone D12 is the initial h1B4 transfectant expression clone, isolated by selection for the GS marker gene by selection in glutamine-free medium. The daughter clone 1G4 was recovered by selection of a D12 population for resistance to 10μM MSX, while its daughter clone, F11, was derived by single-cell dilution cloning at the same MSX level in serum-free medium.

Previous studies had shown that clone D12 contained a single copy of the h1B4 expression vector (Benincasa, D. *et al.*, manuscript in preparation). To evaluate the effect on expression vector copy number of selecting for resistance to MSX, genomic DNA isolated from D12, 1G4 and F11 was digested with XbaI and probed with the human *IGKC* fragment. This fragment will hybridize to the transfected gene and, because it is located close to the restriction enzyme site used to linearize the transfected vector, will identify the restriction fragment that contains the plasmid/mouse genomic DNA junction. This probe identified a 2.9 kb band in D12 genomic DNA representing the single copy of integrated h1B4 Ab expression vector (Figure 1). Signal intensity was greater in DNA from 1G4 and F11. Quantitation of the normalized band intensities indicated that there are 3 and 5 copies of the h1B4 genes in 1G4 and F11, respectively.

The three clones were adapted to suspension culture in serum-free medium, differing only in the presence (clones 1G4 and F11) or absence (clone D12) of MSX. Expression of h1B4 was measured over 3-4 days in suspension culture in the absence of drug, and normalized to cell density. D12, was found to have a specific Ab expression level of 12 pg/c/d, while 1G4 was measured at 43 pg/c/d and F11 at 50 pg/c/d. The increase in Ab expression correlated with the increased vector copy number in these cell lines. 1G4 appeared to grow somewhat more vigorously than F11 and was selected for much of the process research discussed below.

Fed-Batch Culture. When 1G4 was seeded at 2 x 10^5/mL and grown in serum-free batch culture, the culture remained viable for 7 days, reaching a peak density of 1 x 10^6/mL, and secreting 140 mg/L of Ab. Although the apparent doubling time of the culture was 28 hr., growth lagged for one day post-planting and plateaued by the fourth day (Figure 2). Analysis of spent medium from such

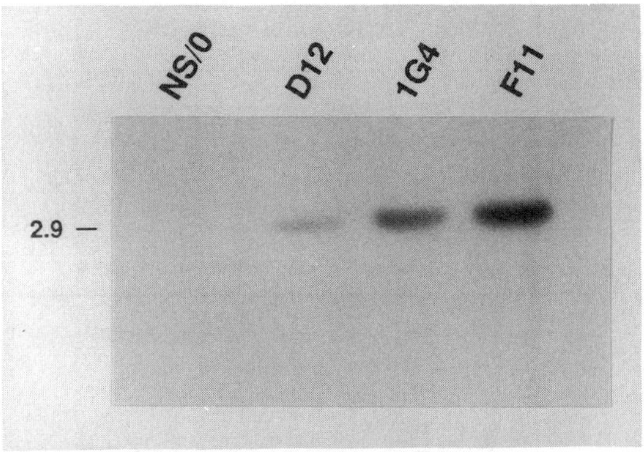

Figure 1. Southern blot of restriction enzyme (XbaI) digest of genomic DNA from wild type NS0 and h1B4 expressing clones D12, 1G4, and F11 hybridized with a probe containing the human Ig kappa constant region gene.

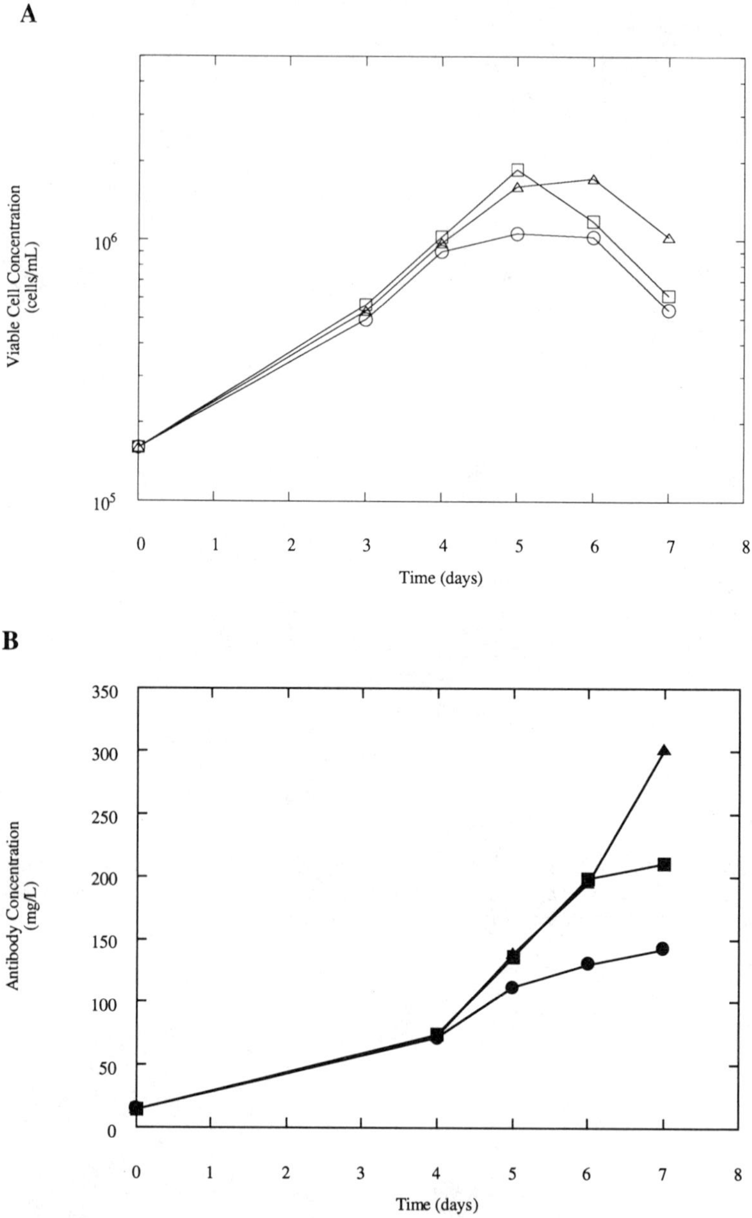

Figure 2. Viable cell (A) and Ab (B) concentration profiles from batch (circles) and two fed-batch cultures of NS0 clone 1G4. Fed-batch 1 (squares) received a 1X addition of a concentrated amino acid solution on day 4 and fed-batch 2 (triangles) received a 2.5X addition of the same concentrated amino acid solution on day 4, as described in the text.

cultures revealed that only 1g/L of glucose had been consumed and only micromolar quantities of ammonia had accumulated. Amino acid analysis of spent culture medium showed that asparagine and glutamic acid were completely consumed by the end of the batch culture (data not shown). This is consistent with their serving as substrates for the glutamine synthetase reaction, asparagine as a source of ammonia, which is combined with glutamate to form glutamine. Simply doubling the initial concentration of these two amino acids in batch culture increased the maximal cell density by 40% to 1.4 x 10^6/ml, prolonged the culture longevity by one day to eight days and increased the Ab titer to slightly over 200 mg/L (data not shown).

A detailed analysis of amino acid metabolism provided a profile of the daily utilization rates (Figure 3), from which a more complete amino acid supplement was formulated. A single addition of this supplement, designed to replenish all spent amino acids, increased the final Ab titer to just 210 mg/L (Fed batch 1 in Figure 2). However, an inadvertent addition of this amino acid supplement to 2.5-fold its intended volume increased the final Ab titer to greater than 300 mg/L (Fed batch 2 in Figure 2). This serendipitous finding suggested that the consumption rate of certain amino acids is concentration dependent. Amino acid analysis on the day post feeding confirmed that the 2.5X fed culture used nearly twice as much asparagine and glutamate as the 1X fed culture (data not shown).

Based on the above preliminary results, a two-feed system was implemented. The most rapidly depleted amino acids were replenished on day 3. Both these and the secondarily depleted amino acids were added on day 6. In this case culture longevity was extended beyond 10 days and Ab titer was increased to 430 mg/L. Again, post-feed depletion analysis allowed further refinement of the nutrient mixture compositions, adjusting for growth and production phase amino acid consumption profiles. A more gradual feeding regime was implemented with daily additions of nutrients. The growth-phase amino acid solution was added on days 3 to 5, glucose on day 6, and the production-phase amino acid solution on days 6 to 8. The cells were now maintained in an apparent stationary plateau phase at near constant specific productivity, 46 pg/c/d. In both shake flasks and bioreactors, similar growth and production profiles were observed and Ab accumulated to over 600 mg/L (data not shown). The further application of this iterative medium depletion analysis approach led to minor refinements of the nutrient feed composition, and Ab titers of over 800 mg/L. A near-homeostatic concentration of most amino acids was now maintained throughout the fed-batch culture. The same feed composition and schedule applied to subclone F11 resulted in Ab accumulations up to 960 mg/L (data not shown).

At this point, it became obvious that nutrients or factors other than glucose and amino acids had become limiting. Empirically, it was determined that proteins, lipids in the form of an emulsion (*49*), ß-mercaptoethanol, ethanolamine and vitamins all contributed to culture longevity and productivity. Addition of these components, amino acids, and glucose in an optimized fed-batch process yielded 1.8 g/L of Ab in a 28 day culture of 1G4. When the same nutrient solutions were fed to F11, the culture declined by day 23 and yielded just 1.2 g/L (Figure 4). Coupled with the finding that each NS0 clone displays a different metabolic profile (*50*), this suggests that feed compositions must be optimized for each cell line.

A general paradigm for an iterative approach to fed-batch process design has emerged from our work as illustrated by this case study. Nutrient utilization is first estimated from batch cultures, and then from simple fed-batch cultures. The objective of continued refinement of daily nutrient feeds is the maintenance of as close as possible to a homeostatic nutrient environment for the cells. By this means, the longevity of the culture may be greatly prolonged, with a constant specific Ab productivity contributing to ever increasing product yields. Once the

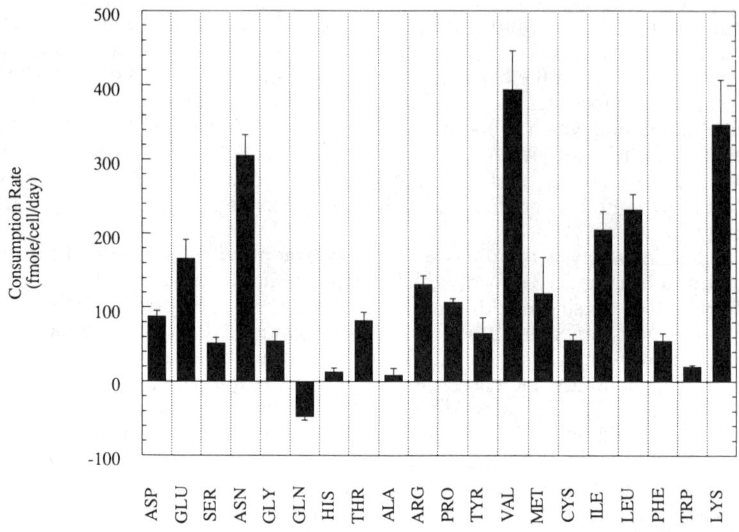

Figure 3. Specific amino acid consumption rate during batch culture of NS0 clone 1G4. Results are averaged from three cultures.

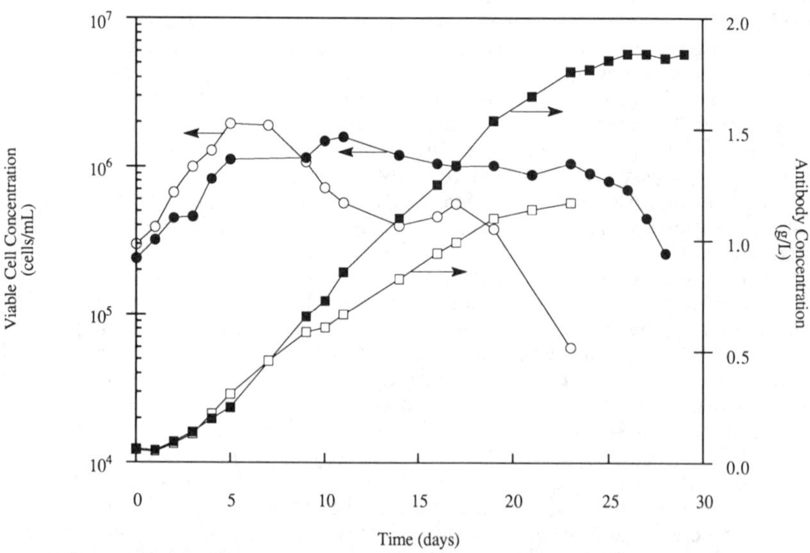

Figure 4. Viable cell (circles) and Ab (squares) concentration profile from an optimized fed-batch culture of NS0 clone 1G4 (closed symbols) and clone F11 (open symbols).

limits of glucose and amino acid feeding are reached, the nature of and timing of addition of other nutrients and factors, representing most of the metabolizable or labile components of the medium, must be empirically determined. Although not particular effective in this case, manipulation of the medium osmolarity or addition of specific inducers, *e.g.* sodium butyrate, can also be applied to increase the specific productivity as well (*35, 51-52*). In our paradigm, this represents the final stage of fed-batch optimization.

Perfusion Culture. The same nutrient utilization data which define the optimization of fed-batch culture may also be used to design efficient perfusion processes. Although most perfusion processes rely on continuous addition of basal medium (for example see *37*), the further addition of nutrient concentrates similar to those used in the fed-batch process can be used to supplement the perfusion of fresh medium. In the example illustrated in Figure 5, a perfusion culture of 1G4, amino acid concentrate was introduced into the reactor on day 5, when the cells exceeded 10^6/mL, just as in fed-batch culture. Perfusion of fresh medium was begun on day 6 at 0.25 reactor volumes per day (v/d). To minimize product dilution, amino acid concentrate was also added to the perfusion reservoir on day 8. On subsequent days, richer, more complex nutrient concentrates were added to both the reactor and the perfusion reservoir. With such concentrate additions, a maximum volumetric perfusion rate of only 1.5 v/d supported cultures at 2 x 10^7/mL (Figure 5) with approximately the same specific productivity as previously measured in fed-batch culture. Ab titers of up to 0.5 g/L were maintained in the harvest stream. Based on the final perfusion rate of 1.5 v/d, volumetric productivities greater than 0.6 g/L/d were reached. In 16 days of culture, nearly 50 g of Ab were produced in this 16 L stirred tank.

Discussion

The expression of Abs as recombinant genes in animal cells allows the application of genetic engineering both to optimize the characteristics of the Ab protein as well as to optimize expression levels. Lymphoid cell lines, such as myeloma cells, appear particularly well suited to high-level recombinant Ab expression. In the case study we have presented, transfection of NS0 cells with a tandem vector containing the GS marker gene and heavy and light chain genes on one plasmid, and selection in glutamine-free medium allowed isolation of clones expressing high levels of Ab. Clone D12 expresses 12 pg/c/d from a single gene copy. By selecting for resistance to MSX, a daughter clone, 1G4, was isolated expressing a 3.5-fold higher level, while single-cell cloning of 1G4 in serum-free medium identified F11, which expresses at a four-fold higher level than D12. These increases in expression correspond well with the estimated relative Ab gene copy numbers, 3 for 1G4 and 5 for F11.

Expression of GS allows growth in glutamine free medium and minimizes ammonia production, making such clones particularly suitable for long-term fed-batch culture. In batch culture, 1G4 produced 140 mg/L of Ab. Through an iterative procedure of refining the amino acid composition of concentrated nutrient feed solutions, and by adding these solutions on a daily basis in fed-batch culture together with glucose, the productive culture lifetime was increased and Ab accumulated to over 800 mg/L. A near homeostasis of rate-limiting nutrients was maintained throughout, and similar or even enhanced specific productivities were preserved. Addition of an empirically derived mixture of extra nutrients extended the culture to 28 days and resulted in a final Ab titer of 1.8 g/L. On the other hand, the identical process, when applied to clone F11, which has a nominally higher specific productivity, saw an earlier culture demise at 23 days and a lower final titer

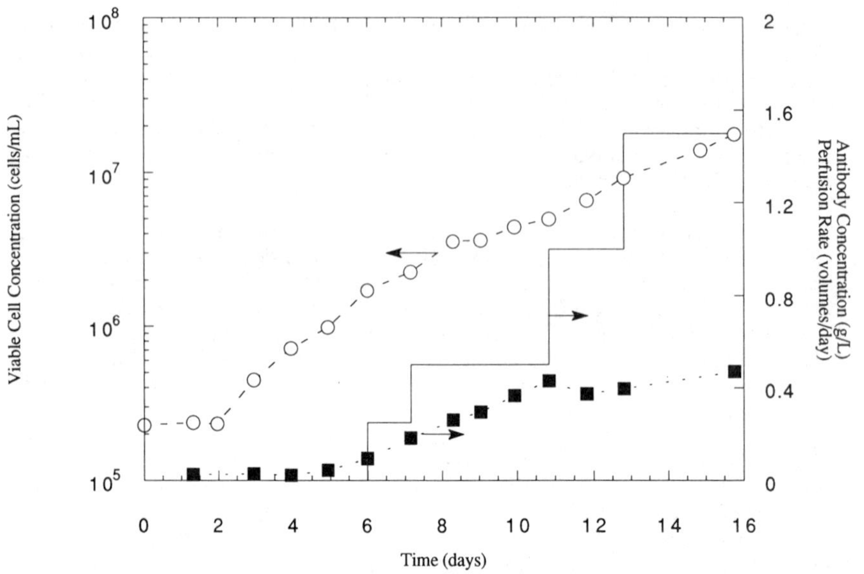

Figure 5. Viable cell (circles) and Ab (squares) concentration profile from a perfusion culture of NS0 clone 1G4. The solid line represents the perfusion rate. Starting on day 5, concentrated nutrient solutions were added to the reactor and from day 8 also to the reservoir as detailed in the text.

of only 1.2 g/L. This underscores the need for customized nutrient feeds for each clone to accommodate individual nutrient metabolic profiles.

Addition of similar nutrient concentrates to a perfusion culture reduced the volumetric perfusion rate required to support high cell densities, up to 2 x 10^7 /mL, and allowed the harvest of Ab at high titers, up to 0.5 g/L. Based on the reactor volume, a maximum volumetric productivity in excess of 0.6 g/L/d was obtained. Thus, by combining customized nutrient concentrate feeding with basal medium perfusion, both high titer and high volumetric throughput can be achieved.

Nonetheless, the relative simplicity of fed-batch cultures is attractive. We, and others, have often speculated as to the maximum Ab titers that can be reached in such cultures. The best cell lines reported to date express up to 100 pg/c/d of Ab, while the highest viable cell concentrations achieved in fed-batch cultures reach 4 x 10^6/mL. Assuming that one could maintain such high expressing cell lines at these high viable cell densities for a period of 2 weeks, final titers could climb above 5 g/L! The limits of such processes have clearly not yet been reached.

Acknowledgements. The authors thank R. Ellis for his support and encouragement and S. Pols for preparation of this manuscript.

Literature Cited

1. Larrick, J. W.; Fry, K. E. *Hum. Antibod. Hybridomas* **1991**, *2*, 172-189.
2. *Antibody Engineering: A Practical Guide*; Borrebaeck, C., Ed.; W.H. Freeman and Co.: New York, NY 1992.
3. Morrison, S. L.; Johnson, M. J.; Herzenberg, L. A.; Oi, V. T. *Proc. Natl. Acad. Sci. (USA)* **1984**, *81*, 6851-6855.

4. Morrison, S. L.; Canfield, S.; Porter, S.; Tan, L. K.; Tao, M.; Wims, L.A. *Clin.Chem.* **1988**, *34*, 1668-1675.
5. Singer, I. I.; Kawka, D. W.; DeMartino, J. A.; Daugherty, B. L.; Elliston, K. O.; Alves, K.; Bush, B. L.; Cameron, P. M.; Cuca, G. C.; Davies, P.; Forrest, M. J.; Kazazis, D. M.; Law, M. -F.; Lenny, A. B.; MacIntyre, D. E.; Meurer, R.; Padlan, E. A.; Pandya, S.; Schmidt, J. A.; Seamans, T. C.; Scott, S.; Silberklang, M.; Williamson, A. R.; Mark, G. E. *J. Immunol.* **1993**, *150*, 2844-2857.
6. Roberts, S.; Cheetham, J. C.; Rees, A. R. *Nature* **1987**, *328*, 730-734.
7. Huse, W.; Sastry, L,; Iverson, S. A.; Kang, A. S.; Alting-Mees, M.; Burton, D. R.; Benkovic, S. J.; Lerner, R. A. *Science* **1989**, *246*, 1275-1281.
8. Ward, E. S.; Gussow, D.; Griffith, A. D.; Jones, P. T.; Winter, G. *Nature* **1989**, *341*, 544-546.
9. Weidle, U. H.; Borgya, A.; Mattes, R.; Lenz, H.; Buckel, P. *Gene* **1987**, *51*, 21-29.
10. Page, M. J.; Sydenham, M. A. *Biotechnology* **1991**, *9*, 64-68.
11. Fouser, L. A.; Swanberg, S. L.; Lin B.-Y.; Benedict, M.; Kellerher, K.; Cumming, D. A.; Reidel, G. E. *Biotechnology* **1992**, *10*, 1121-1127.
12. Schulman, M.; Wilde, C.D.; Köhler, G. *Nature* **1978**, *276*, 269-270.
13. Cowan, N. J.; Secher, D. S.; Milstein, L. *J. Mol. Biol.* **1974**, *90*, 691-701.
14. Köhler, G. *Proc. Natl. Acad. Sci. (USA)* **1980**, *77*, 2197-2199.
15. Dorai, H.; Moore, G. P.; *J. Immunology* **1987**, *139*, 4232-4241.
16. Sawai, H.; Yamasaki, N.; Komori, S.; Karasuyama, H.; Isojima, S. *Am. J. Reprod. Imunol.* **1992**, *29*, 100-108.
17. Arakawa, F.; Haruno, M.; Kuroki, M.; Kanda, H.; Watanabe, T.; Misumi, Y.; Matsuoka, Y. *Hybridoma* **1993**, *12*, 365-379.
18. Crowe, J. S.; Hall, V. S.; Smith, M. A.; Cooper, H. J.; Tite, J. P. *Clin. Exp. Immunol.* **1992**, *87*, 105-110.
19. Neumaier, M.; Shively, L.; Chen, F.-S.; Guida, F.-J.; Ilgen, C; Paxton, R. J.; Shively, J. E.; Riggs, A. D. *Cancer Research* **1990**, *50*, 2128-2134.
20. Heinrich, G.; Gram, H.; Kocher, H. P.; Schreier, M. H.; Ryffel, B.; Akbar, A.; Amlot, P. L.; Janossy, G. *J. Immunology* **1989**, *143*, 3589-3597.
21. Beidler, C. B.; Ludwig, J. R.; Cardenas, J.; Phelps, J.; Papworth, C. G.; Melcher, E.; Sierzega, M.; Myers, L. J.; Unger, B. W.; Fisher, M.; David, G. S.; Johnson, M. J. *J. Immunology* **1988**, *141*, 4053-4060.
22. Shu, L.; Qi, C.-F.; Schlom, J.; Kashmiri, S. V. S. *Proc. Natl. Acad. Sci. USA* **1993**, *90*, 7995-7999.
23. Shitara, K.; Nakamura, K.; Tokutake-Tanaka, Y,; Fukushima, M.; Hanai, N. *J. Immunological Methods* **1994**, *167*, 271-278.
24. Bebbington, C. R.; Renner, G.; Thomson, S.; King, D.; Abrams, D.; Yarranton, G. T. *Biotechnology* **1992**, *10*, 169-175.
25. Robinson, D. K.; Seamans, T.C.; Gould, S.L.; DiStefano, D.J.; Chan, C.P.; Lee, D.K.; Bibila, T.; Glazomitsky, K.; Munshi, S.; Daugherty, B.; O'Neill Palladino, L.; Stafford-Hollis, J.; Hollis, G. F.; Silberklang, M. *Annals N. Y. Academy Sciences* **1994**, *745*, 285-296.
26. Simonsen, C. C.; McGrogan, M. *Biologicals* **1994,** *22*, 85-94.
27. Savinell, J. M.; Lee, G. M.; Palsson, B. O. *Bioprocess Engineering* **1989**, *4,* 231-234.
28. Wurm, F. M. *Biologicals* **1990**, *18*, 159-164.
29. Baker, M. D.; Shulman, M. J. *Mol. Cell Biol.* **1988**, *8*, 4041-4047.
30. Baker, M. D.; Karn, H. A.; Read, L. R. *J. Immunological Methods* **1994**, *168*, 25-32.
31. Duval, D.; Demangel, C.; Munier-Jolain, K.; Miossec, S.; Geahel, I. *Biotechnol. Bioeng.* **1991**, *38*, 561-570.

32. Duval, D.; Demangel, C.; Miossec, S.; Geahel, I. *Hybridoma* **1992**, *11*, 311-322.
33. Reuveny, S.; Velez, D.; Miller, L.; Macmillan, J. D. *J. Immunological Methods* **1986**, *86,* 61-69.
34. Luan, Y. T.; Matharasan, R.; Magee, W. E. *Biotechnology Letters* **1987**, *9*, 691-696.
35. Bibila, T. A.; Robinson, D. K. *Biotechnology Progress* **1995**, *11*, 1-13.
36. Maiorella, B. In *Harnessing Biotechnology for the 21st Century: Proceedings of the Ninth International Biotechnology Symposium, Crystal City, VA, August 19-21, 1992*; Ladisch, M.; Bose, A., Eds.; American Chemical Society: Salem, MA. 1992; 26-29.
37. Robinson, D. K.; Widmer, J.; Memmert, K. J. *Biotechnology* **1992**, *22*, 41-50.
38. Himmelfarb, P.; Thayer, P. S.; Martin, H. E. *Science* **1969**, *164*, 555-557.
39. Wright, S. D.; Rao, P. E.; Van Voorhis, W.; Craigmyle, L. S.; Iida, K.; Talle, M. A.; Westberg, E. F.; Goldstein, G.; Silverstein, S. C. *Proc Natl. Acad. Sci. (USA)* **1983**, *80*, 5699-5703.
40. Tonnesen, M. G.; Anderson, D. C.; Springer, T. A.; Knedler, A.; Avdi, N.; Henson, P. M. *J. Clin. Invest.* **1989**, *83*, 637-646.
41. Lo, S. K.; Van Seventer, G.; Levin, S. M.; Wright, S. D. *J. Immunol.* **1989**, 143, 3325-3329.
42. Lo, S. K.; Detmers, P. A.; Levin, S. M.; Wright, S. D. *J. Exp. Med.* **1989**, 169, 1779-1793.
43. Daugherty, B.L.; DeMartino, J.A.; Law, M.F.; Kawka, D.W.; Singer, I.I.; Mark, G.E. *Nucleic Acids Research* **1991**, *19*, 2471-2476.
44. DeMartino, J. A.; Daugherty, B.L.; Law, M.F.; Cuca, G. C.; Alves, K.; Silberklang, M.; Mark, G. E. *Antibody Immunoconj. Radiopharmaceut.* **1991**, *4*, 829-835.
45. Bebbington, C. R.; Hentschel, C. C. G. in *DNA cloning: a practical approach, v. 3*; Glover, D. Ed.; IRL Press Ltd: Oxford, U.K. 1987; 163-188.
46. Southern. E. *J. Mol. Biol.* **1975**, *98*, 503-515.
47. Mather, J. *Methods Enzymology*, **1990**, *185*, 157-167.
48. Krips, D. M.; Sitrin, R. O.; Oliver, C. N. *FASEB J* **1991**, *5*, A465.
49. Seamans, T. C.; Gould, S. L.; DiStefano, D. J.; Silberklang, M.; Robinson, D. K. *Annals N. Y. Academy Sciences* **1994**, *745*, 240-243.
50. Gould, S. L.; DiStefano, D. J.; Cuca, G.; Robinson, D. K.; Silberklang, M. *In Vitro Cell. Dev. Biol.* **1992**, *28*, 162A.
51. Oh, S. K. W.; Vig, P.; Chua, F.; Teo, W. K.; Yap, G. J. *Biotech. Bioeng.* **1993**, *42*, 601-610.
52. Bibila, T. A.; Ranucci, C. S.; Glazomitsky, K.; Buckland, B. C.; Aunins, J. G. *Annals N.Y. Academy Sciences* **1994**, *745*, 277-284.

RECEIVED March 27, 1995

Chapter 2

Antibody Production in Human Hybridomas with Growth-Associated Production Kinetics

Hidekazu Sawada[1] and Kazuaki Kitano[2]

[1]Biotechnology Department, Pharmaceutical Research Division and [2]Discovery Research Laboratories II, Takeda Chemical Industries, Ltd., 17–85, Juso-Honmachi 2-chome, Yodogawa-ku, Osaka 532, Japan

Murine monoclonal antibodies (MoAbs) have been widely used for the detection and purification of biologically active substances, and as diagnostic agents for various diseases. For clinical use, however, murine MoAbs are not desirable because of the antigenicity in humans and human MoAbs are preferable.
Hepatitis B virus (HBV) infection is one of the major public health problems in the world. Vaccines have been developed using recombinant DNA technology (*1*) but human MoAbs are also indispensable, because a combination of passive and active immunization was found to be effective for post-exposure prophylaxis (2).
The human-human (h-h) hybridomas system has become an effective technique for preparing human MoAbs. Using h-h hybridomas, we have recently developed an efficient process for industrial production of human MoAbs against hepatitis B virus surface antigen (HBsAg) for practical use in patients with HBV (*3,4*).
This article is an overview of our research for anti-HBsAg human MoAb production. It includes the establishment of h-h hybridomas, improvement in the cell lines, development of serum-free media, the kinetics of human MoAb production, efficient large-scale production process, and the neutralizing activity in chimpanzees of human MoAbs thus produced.

Human Hybridomas Secreting Anti-HBsAg MoAbs

Several experimental systems have been developed to obtain cell lines secreting human MoAbs. Human B lymphocytes transformed with Epstein-Barr virus (EBV) were the first cell lines established to obtain antigen-specific human MoAbs (*5*). However, antibody production by the EBV transformants tends to be short-lived and the production capacity of the cells is generally low (*6*). In the h-h hybridoma systems originally reported in 1980 (*7,8*), the frequency of hybridoma formation was quite low due to the absence of a suitable malignant fusion partner (*9,10*). Therefore, cell fusion between mouse myeloma with high fusion frequency and human peripheral blood lymphocytes (PBL) was attempted to establish producers of human MoAbs (*11,12*). However, it was generally difficult to obtain stable

0097–6156/95/0604–0015$12.00/0

clones due to the preferential loss of human chromosomes in such heterohybridomas. Teng et. al., (*13*) used a mouse-human (m-h) heteromyeloma as a parental cell line to obtain stable producers of human MoAb at a relative high frequency.

In 1985, Ichimori et. al., (*14*) established a mouse.human-human (m.h-h) trioma system to obtain cell lines secreting human MoAbs against tetanus toxoid (T.T.) and HBsAg. In this system the murine myeloma cell line P3-X63-Ag8.U1 (P3U1) was first fused with human PBL, and two parental heterohybrids, HM-3 and HM-5, which grow rapidly and do not produce MoAb were selected as 8-azaguanine-resistant variants. By fusing HM-3 or HM-5 with human PBL from anti-HBsAg or anti-T.T. antibody-positive donors, m.h-h triomas secreting anti-HBsAg or anti-T.T. human MoAbs were obtained. Most of these triomas stably secreted human MoAb, but some were unstable and ceased to produce antibody three or four months after isolation.

As a superior parental cell line that gives a high fusion frequency and yields stable hybridomas that secrets human MoAbs, Ichimori et. al., (*15*) established a 6-thioguanine- and ouabain-resistant cell line, TAW-925, from a 6-thioguanine-resistant variant of the human B lymphoblastoid cell line WI-L2 (*16*). TAW-925 showed a remarkably high fusion frequency of $1.6\text{-}6.2 \times 10^{-5}$ when fused with EBV-transformants. By fusing EBV-transformed human PBL with TAW-925, five h-h hybridomas secreting anti-HBsAg human MoAbs were established, of which four produced specific IgG antibody and one produced specific IgM antibody (*3*). Antibody production by all of these hybridomas was quite stable, even after several transfer for more than one year, but the level of production (2.6-7.5μg/ml) was not high (Table I). MoAbs produced by these hybridomas reacted with all the HBsAg subtypes in test serd including adr, adw, ayr, ayw, adrw, and adyr indicating recognition of the "a" domain common to all known subtypes of HBV (*17*).

Table I: Human-human hybridomas secreting anti-HBsAg MoAbs

Hybridoma	Immunoglobulin class and subclass	Amount of MoAb accumulated	Stability of MoAb production
W471-7.24	IgG_1	2.6 μg/ml	>12 months
HBW-3.7	IgM	3.0 μg/ml	>12 months
HBW-4.16	IgG_1	7.5 μg/ml	>12 months
HBW-6.20	IgG_1	6.1 μg/ml	>12 months
HBW-7.10	IgG_1	4.5 μg/ml	N.T.*

*Not tested. Reprinted from Harada et. al., (1989) by permission of BIO/TECHNOLOGY Publisher

Improvement of Cell Lines

For industrial production of human MoAbs, it was necessary to improve the production capacity of h-h hybridomas. The application of gene technology to the improvement of cell lines has not been very successful except for the case of genetically engineered cells. Onodera et. al., (*18*) succeeded in improving human interleukin-2(IL-2) productivity of Jarkat cells by introducing the E1a gene of adenovirus type 12. Iwamoto et. al. (*19*) introduced the Tac gene into h-h hybridoma to enhance the stimulatory effect of IL-2 on MoAb production, because

IL-2 was found to slightly stimulate production, but the IL-2 receptor (Tac antigen) was not detected on the surface of these hybridomas. Tac gene transfectants produced more than six times as much human MoAb as the control upon addition of IL-2 to the medium. The addition of IL-2 however, was not practical for industrial application.

Another example of our studies is the introduction of the basic fibroblast growth factor (FGF) gene into h-h hybridomas, based on the finding that basic and acidic FGFs markedly stimulated the hybridoma growth from low inoculum density such as 10^3 cells/ml (*20*). The cell growth of basic FGF gene transfectants from a small inoculum was clearly stimulated as was expected and antibody production was also increased. By using these cell lines, we succeeded in shortening seed steps for large scale serum free culture.

The conventional method of repeated recloning and selection was not necessarily effective in h-h hybridomas, probably reflecting the very stable production of the antibody, although high producers were efficiently obtained by this method in the case of a m.h-h heterohybridoma (*21*).

We tried several other approaches to obtain high producers applicable to industrial production and noticed that adaptation of cells to certain stress conditions followed by recloning and selection was an effective method for obtaining high producers in h-h hybridomas. For example, when h-h hybridoma HBW-4.16 was adapted to protein-free medium through stepwise reductions in the concentration of proteinaceous ingredients in the medium, the antibody production increased with each step. The strain HPO-7 obtained after recloning the adapted cells was capable of producing about six times as much antibody as the original cell line (Figure 1). As the result of further adaptation to the same protein-free medium to allow growth from a small inoculum, TFC-7 with increased productivity was also obtained. However, nonspecific IgM, probably originating from the parental TAW-925, was increased in accordance with the increase in the IgG titer. Therefore, we attempted to select non-secretors of IgM by recloning and selection from HPO-7, expecting a further increase in IgG productivity. A non-secretor of IgM, HPO-75, was successfully obtained but IgG productivity did not increase. By adapting HPO-75 to high-speed agitation, we obtained a further improved cell line HPO-75.29AGC-1, which produced 10 times as much MoAb as the original cell line. This cell line could grow very well in a large scale fermentor and produce a large amount of the antibody.

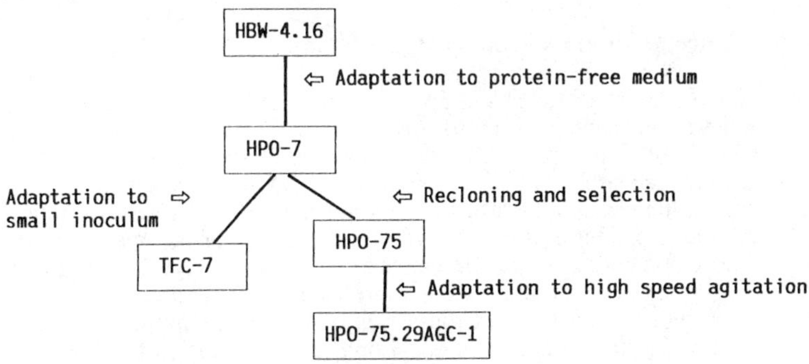

Figure 1: Derivation of high productivity subclones of human hybridoma HBW-4.16 secreting anti-HBsAg human MoAb.

The other high producing subclones WSO-20.68 and HBW-6.129 were also derived from the original strains W471-7.24 and HBW-6.20 respectively.

Serum-Free Media for Human Hybridomas

Serum-containing media have been most commonly used for mammalian cell culture, but they have several disadvantages for industrial application. Therefore, it is very important to establish chemically defined serum-free media composed of inexpensive compounds for industrial production of human MoAbs. Since Sato (*22*) proposed that the major role of serum in cell culture is to provide hormones, the hormonal requirements of various cell lines have been investigated and chemically defined serum-free media have been developed for many cell lines. (*23*)

Shintani et. al., (*24*) established chemically defined serum-free media containing polyethylene glycol (PEG) which were effective for not only hybridomas but also for a variety of lymphoid cells. A serum-free medium, PEG-86-1, which contains all the ingredients of ITL-2 medium and 0.1%(W/V) PEG 20,000 was useful for h-h hybridomas producing human MoAbs against HBsAg. The ITl-2 medium contains 2mg/L insulin, 2mg/L transferrin, 2 x 10 $^{-6}$M ethanolamine and 2.5 x 10 $^{-8}$M sodium selenite (ITES) (*25*), 4.5g/L HEPES and 1g/L sodium bicarbonate in a mixture (1:1:2) of IMDM, Ham F12 and L-15 media.

Growth-Associated Production of Human MoAbs

Perfusion culture in which the spent medium is continuously replaced with fresh medium is one effective technique for cultivating hybridoma cells for a long period and of obtaining a high cell density and high productivity (*26, 27*). However, when the h-h hybridoma TFC-7 was cultivated in PEG-86-1 medium using this system, antibody production ceased in accordance with stoppage of growth. Active synthesis of human MoAb occurred at the beginning of the culture but production dropped rapidly day by day after the time when the cell number arrived at an almost constant level, even though viable cell count was maintained (*28*). With regard to other h-h hybridomas secreting anti-HBsAg human MoAbs shown in Table I, the production patterns in perfusion culture were the same. There were also no changes when other media were used in place of PEG-86-1. In semi-continuos culture without perfusion, the antibody was produced repeatedly in accordance with repeated cell growth. Thus, the synthesis of human MoAb in our h-h hybridomas proved to be associated with cell growth.

Efficient Culture Method for Human Hybridomas

To search for the most efficient culture method for h-h hybridomas, we tested various methods of suspension culture using a 2-L jar fermentor and of immobilized culture using a ceramic matrix and a hollow fiber reactor (*28*). For these experiments, the strain TFC-7 and PEG-86-1 medium were used.

Chemostat culture was not useful for this hybridoma which has a low growth rate. To maintain productivity effectively for a long period, we developed a continuous-perfusion culture system in which chemostat and perfusion culture were combined (Figure 2). This system, both culture supernatant and culture broth were continuously harvested and fresh medium was continuously supplied at the same rate as the total of perfusion and dilution. Long-term cultivation for over 50 days was achieved, and the average productivity per unit working volume per day was higher , but the productivity per unit volume of spent medium was much lower than that for batch culture.

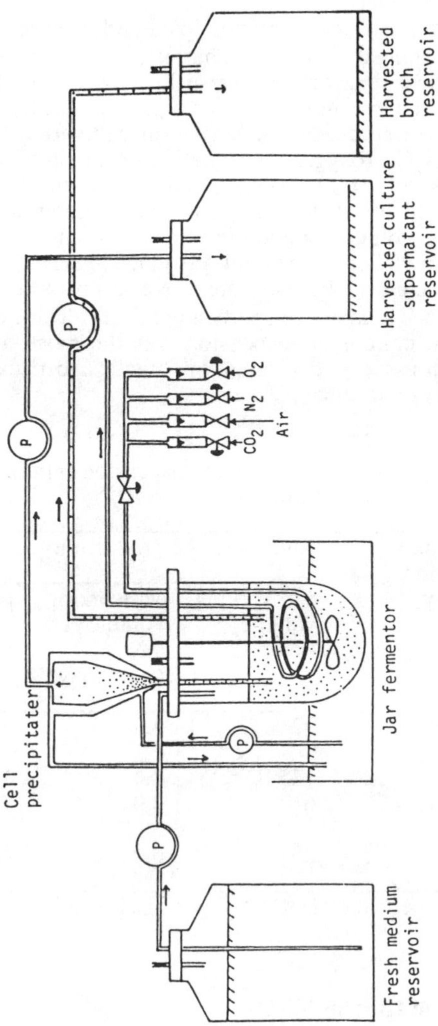

Figure 2: A continuous-perfusion culture system devised for a 2-L jar fermentor. Reprinted from Shintani et. al., (1991) by permission of Kluwer Academic Publishers.

A short-term perfusion culture was successfully performed by increasing the perfusion rate [feed volume of fresh medium (=harvest volume of culture supernatant)/working volume/day] to 1.5 day $^{-1}$. The cell density reached about 1 x 10^7 cells/ml resulting in high production. The productivity was higher than that of the continuous-perfusion culture described above based on both per unit working volume per day and per unit volume of spent medium,. although the duration of culture was only 17 days (Table II).

When the hybridoma was cultivated in immobilized-perfusion reactors such as a ceramic matrix reactor (Opticell TM5200R, Charles River Biochemical Services Inc., MA, USA) or a hollow fiber reactor (Tabai Espec Co., Inc., Osaka, Japan), antibody production did not continue for a long period, similarly to perfusion culture in suspension. We succeeded in long-term cultivation by removing a portion of the cells when OCR (oxygen consumption rate; monitor of cell growth) reached a plateau. This culture method, termed semicontinuous immobilized perfusion culture, was the most efficient for the production based on the productivity per working volume (immobilized bed volume) per day, but the productivity per unit volume of spent medium was much lower than that for short-term perfused-suspension culture. Furthermore, there seems to be some problems in scaling up semicontinuous immobilized perfusion culture. Thus, we concluded that short-term perfusion culture in suspension was the most suitable for h-h hybridomas with growth-associated production kinetics, because of the high productivity and simplicity of scaling up.

Table II: Human MoAb productivity in various suspension cultures using 2-L jar fermentor

Culture method	Culture period (day)	Human MoAb productivity		
		per working volume (mg/L)	per working volume per day (mg/L/day)	per spent medium (mg/L)
Batch culture	8	32	4.0	32
Short-term perfusion culture	17	125	7.4	13
Chemostat culture	31	61	2.0	7.0
Continuous perfusion culture	50	252	5.0	4.2

Reprinted from Shintani et. al., (1991) by permission of Kluwer Academic Publishers

Large-Scale Production of Human MoAbs

We scaled up the short-term perfusion culture from a 2-L jar fermentor to a 50-L fermentor (working volume, 25L) equipped with porous Teflon tubes (Sumitomo Electric Co., Japan) for aeration, rotating shear filter (Dynamic Biopressure Filter; Sulzer AG, Switzerland) for separation of supernatant from the culture broth, fresh medium reservoir and harvested culture-supernatant reservoir (Figure 3; *4*). The most improved cell line, HPO-75.29AGC-1, and serum-free PEG-86-1 medium were used in the culture. Perfusion was initiated from the fourth day of culture, and

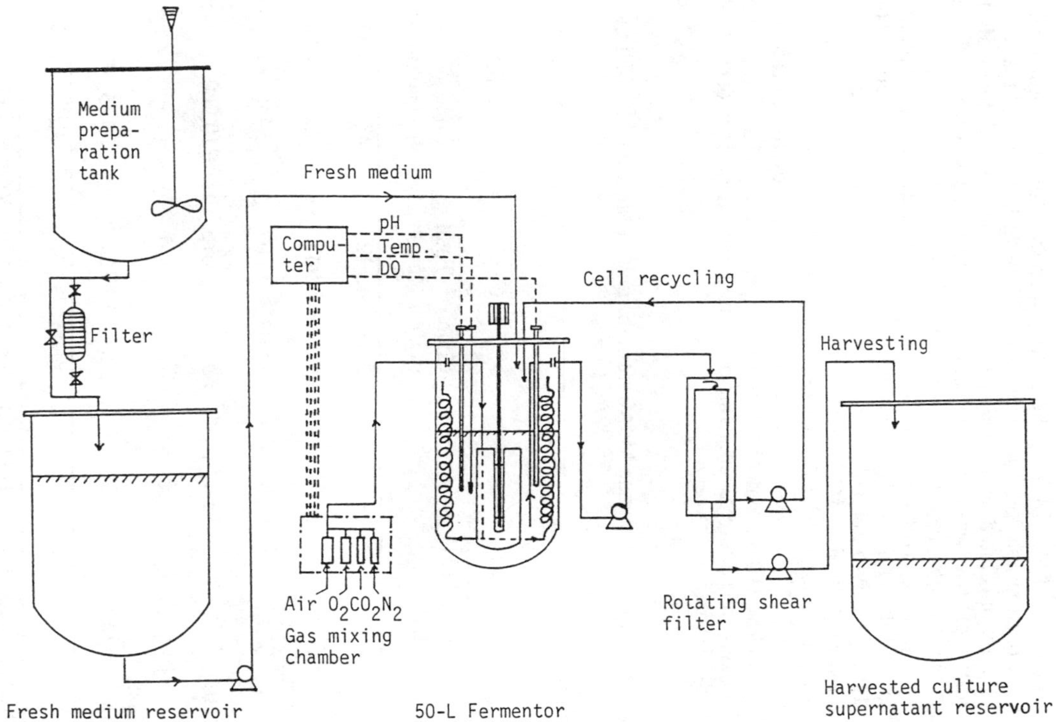

Figure 3: A perfusion culture system devised for a 50-L fermentor. Reprinted from Sawada et. al., (1995) by permission of springer-Verlag, GmbH & Co., KG.

the maximum perfusion rate was controlled at 1.0 day $^{-1}$. The hybridoma proliferated to a cell density of about 1 x 10^7 cells/ml, and a total of about 6.3g of human MoAb was produced in perfusion culture for 15 days with controlled pH (7.2) and dissolved oxygen (about 1ppm) (Figure 4). The MoAb productivity was 253mg per L of working volume, 16.9mg per L of working volume per day, 26mg per L of spent medium.

For purification of the MoAb, the culture supernatant was subjected to hydrophobic chromatography on Butyl-Cellulofine (Seikagaku Kogyo, Japan). The crude MoAb fraction eluted with 10%-saturated ammonium sulfate in 50mM phosphate buffer (pH 6.8) was concentrated, desalted and successively purified by ion-exchange chromatography on CM-Toyopearl (Tosoh, Japan), high-performance liquid chromatography (HPLC) on Ceramic hydroxyapatite (Mitsui Toh-atsu, Japan) and gel filtration on Toyopearl HW-55F (Tosoh). The pooled main fraction was passed through a Pyro-Sep column (Daicel, Japan) and then filtered through a Millex membrane (Millipore, MA, USA) to afford the final bulk solution of MoAb. The yield was approximately 75% for the entire purification procedure using HBW4 antibody (HPO-75.29AGC-1 culture supernatant) as an example (TableIII).

The other MoAbs, HBW6 and W471, were also prepared on a large scale in the same manner using the culture supernatants of the high producing subclones HBW-6.129 and WSO-20.68, respectively.

Table III: Purification of human MoAb from the culture supernatant of HPO-75.29AGC-1.

Step	Volume (L)	A_{280}	IgG			
			GPC		ELISA	
			Conc. (μg/ml)	Total (mg)	Conc. (μg/ml)	Total (mg)
Culture supernatant	125	----	32.9	4,112	43	5,375
Butyl-cellulofine eluate	7.5	0.928	526	3,945	626	4,695
CM-Toyopearl eluate	1.0	5.00	3,807	3,807	4,720	4,720
Hydroxyapatite eluate	0.39	13.47	9,428	3,676	10,960	4,274
Toyopearl HW-55F eluate	1.835	2.63	1,875	3,440	2,012	3,692
Anti-HBsAg	3.675	1.213	887	3,259	1,125	4,134
			Yield: 79.3%		Yield: 76.9%	

The purified preparations of HBW4, HBW6, and W471 were judged to be homogeneous on sodium dodesyl sulfate-polyacrylamide gel electrophoresis (SDS-PAGE) (Figure 5), and exhibited a single peak on gel-permeation chromatography using TSK gel G3000SW (Tosoh). These were also confirmed to be well-suited for therapeutic purposes (*4*).

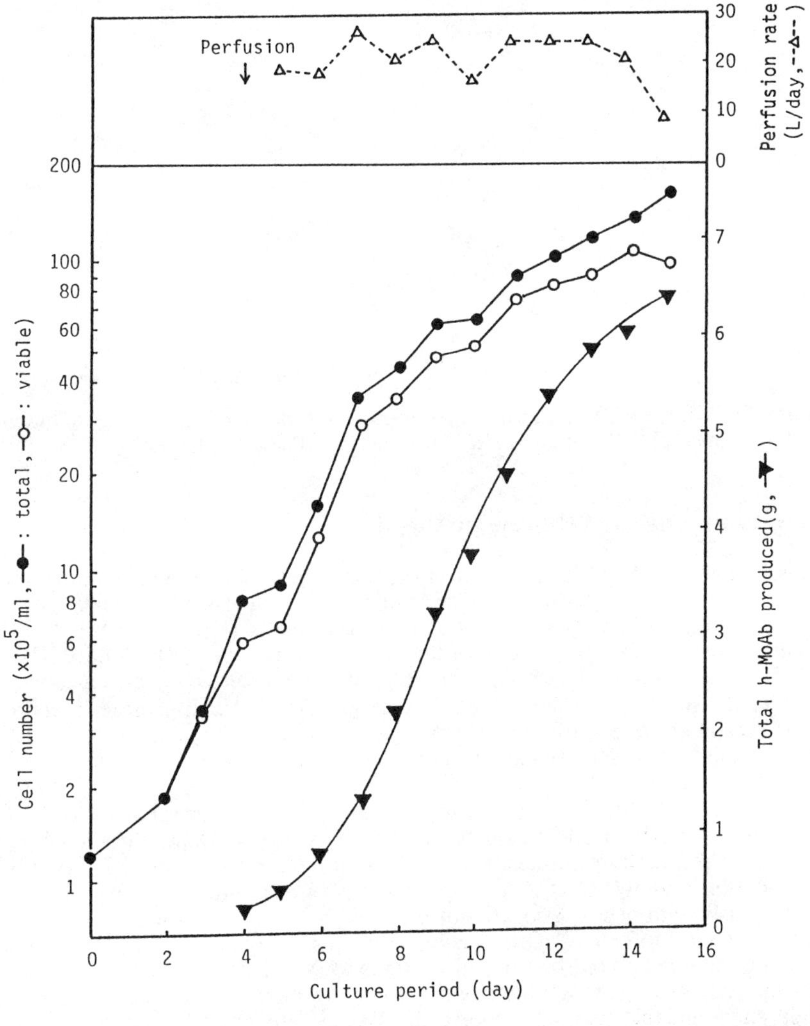

Figure 4: Typical time courses of hybridoma proliferation and human MoAb production in short-term perfusion culture in a 50-L fermentor. HPO-75.29AGC-1 cells were inoculated at $1.2x10^5$ cells/ml into 25 L of serum-free PEG-86-1 medium and cultivated at 35°C at an agitation speed of 40rpm. Dissolved oxygen and pH were controlled at around 1ppm and pH 7.2, respectively. Perfusion was begun from the fourth day of the culture. Reprinted from Sawada et. al., (1995) by permission of springer-Verlag GmbH & Co. KG

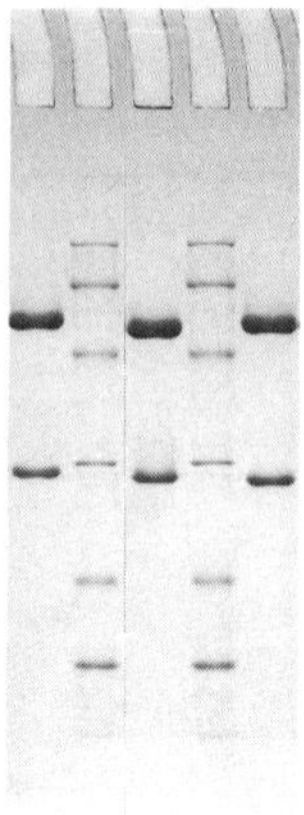

Figure 5: SDS-PAGE pattern in reducing conditions of anti-HBsAb human MoAbs. Lane 1: HBW4, Lane 2: W471, Lane 3: HBW6, m: marker

Neutralizing Activity of Human MoAbs

Neutralizing tests of anti-HBsAg human MoAbs were performed with chimpanzees at the TNO Primate Center in the Netherlands. In the first test, an antibody cocktail (6 mg) composed of an equal amount of three IgG MoAbs (HBW4, HBW6, W471) prepared from culture supernatants of the original hybridomas (HBW-4.16, HBW-6.20, W471-7.24) was injected intravenously into a chimpanzee after preincubation at 37°C for 60 minutes with infectious ayw HBV[10^3 chimpanzee-infectious dose(CID_{50})]. One control animal was challenged with the same dose of HBV alone. As judged by serological results for HBV-associated hepatitis, the antibody cocktail neutralized HBV infectivity (*3*).

In the second test, five chimpanzees were challenged intravenously with 10^3 CID_{50} of ayw HBV. Human MoAbs(HBW4, HBW6, W471) were prepared on a large scale from culture supernatants of the improved strains HPO-75.29AGC-1, HBW-6.129 and WSO-20.68, respectively. Four animals were injected intravenously with one or a cocktail of the three MoAbs (15 mg per animal) three hours after viral infection. One was employed as the control. All four animals receiving the human MoAbs showed no signs of viral infection up to six months after the challenge, although the control animal developed acute hepatitis. These results suggested that the human MoAbs HBW4, HBW6, and W471, produced on a large scale from the improved strains, alone or in combination, protected chimpanzees against the development of hepatitis B (*4*).

Conclusions

Ichimori et. al., (*15*) developed an excellent h-h hybridoma system to establish stable cell lines secreting human MoAbs. Shintani et. al., (*24*), found that PEG stimulates the growth of a wide variety of cell lines under serum-free or low serum conditions. A chemically defined serum-free medium, PEG-86-1, was applicable for h-h hybridomas producing anti-HBsAg MoAbs. Based on these basic technologies developed by our co-workers, we have established an efficient industrial process for production of human MoAbs for clinical use in patients with HBV.

As the productivity of human MoAb producers was quite low, we investigated the possibility of improving the productivity of the hybridomas and found that adaptation of cell to certain stress conditions such as protein-free medium and high-speed agitation followed by recloning was an effective method. As a result, an improved cell line HPO-75.29AGC-1, which produces 10 times as much MoAb as the original cell line and grows very well in a large-scale fermentor, was successfully obtained.

Growth-associated production is characteristic of these h-h hybridomas. Long-term perfusion culture, in which viable cells are maintained at an almost constant level without cell growth for a long period, was not efficient, because active synthesis of human MoAb only occurred during cell proliferation. As a result of comparison of various culture methods, we decided that short-term perfusion culture should be a practical method for h-h hybridomas with growth-associated production kinetics.

Large-scale cultivation of the cell line HPO-75.29AGC-1 was performed under short-term perfusion using a 50-L fermentor, and the purification of IgG-MoAb was efficiently carried out in yields of approximately 75% on the scale of more than 100L of culture supernatant.

These IgG MoAbs neutralized HBV infectivity and protected chimpanzees against the development of hepatitis B.

Human MoAbs investigated here are expected to contribute to human welfare in the near future.

Literature Cited

1. Itoh, Y. and Fujisawa, Y. (1986) Synthesis in yeast of hepatitis B virus surface antigen modified P31 particles by gene modification. Biochem. Biophys. Res. Commun., 141, 942-948.
2. Szmuness, W., Stevens, C.E., Oleszko, W.R., Goodman, A. (1981) Passive-active immunization against hepatitis B: Immunogenicity studies in adult americans. Lacet 1, 575-577.
3. Harada, K., Ichimori, Y., Sasano, K., Sasai, S., Kitano, K., Iwasa, S., Tsukamoto, K., and Sugino, Y., (1989) Human-human hybridomas secreting hepatitis B virus-neutralizing antibodies. Bio/Technology, 7, 374-377.
4. Sawada, H., Iwasa, S., Nishimura, O., and Kitano, K. (1995) Efficient production of anti-hepatitis B virus antibodies and their neutralizing activity in chimpanzees. Appl. Microbiol. Biotechnol., in press.
5. Steinitz, M. Klein, G., Koskimies, S., and Makel, O. (1977) EB virus induced B lymphocyte cell lines producing specific antibody. Nature, 269, 420-422.
6. Kozbor, D. and Roder, J.C. (1981) Requirements for the establishment of high-titered human monoclonal antiboides against tetanus toxoid using the Epstein-Barr virus technique. J. Immunol., 127, 1275-1280.
7. Olsson, K. and Kaplan, H.S. (1980) Human-human hybridomas producing monoclonal antibodies of predifined antigenic specificity. Proc. Natl. Acad. Sci., U.S.A., 77, 5429-5431.
8. Croce, C.M., Linnenbach, A., Hall, W., Steplewski, Z., and Koprowski, H., (1980) Production of human hybridomas secreting antibodies to measles virus. Nature, 288, 488-489.
9. Edwards, P.A.W., Smith, C.M., Neville, A.M., and O'Hare, M.J., (1982) A human-human hybridoma system based on a fast growing mutant of the ARH-77 plasma cell leukemia derived line. Eur. J. Immunol., 12, 641-648.
10. Cote, R.J., Morrissey, D.M., Houghton, A.N., Beattle, E.J., Oettgen, H.F., and Old, L.J., (1983) Generation of human monoclonal antibodies reactive with cellular antigens. Proc. Natl. Acad. Sci., U.S.A., 80, 2026-2030.

11. Nowinski, R., Berglund, C., Lane, J., Lostrom, M., Bernstein, I., Young, W., Hakomori, S., Hill, L., and Cooney, M. (1980) Human monoclonal antibody against Forssman antigen. Science, 210, 537-539.
12. Scholm, J., Wunderlich, D., and Teramoto, Y.A. (1980) Generation of human monoclonal antibodies reactive with human mammary carcinoma cells. Proc. Natl. Acad. Sci., U.S.A., 77, 6841-6845.
13. Teng, N.N.H., Lam, K.S., Riera, F.C., and Kaplan, H.S. (1983) Construction and testing of mouse-human heteromyelomas for human monoclonal antibody production. Proc. Natl. Acad. Sci., U.S.A., 80, 7308-7312.
14. Ichimori, Y., Sasano, K., Ithoh, H., Hitotsumachi, S., Kimura, Y., Kaneko, K., Kida, M., and Tshukamoto, K. (1985) Establishment of hybrodoma secreting human monoclonal antibiodies against tetanus toxin and hepatitis B virus surface antigen. Biochem. Biophys. Res. Commun., 129, 26-33.
15. Ichimori, Y., Harada, K., Hitotsumachi, S., and Tsukamoto, K. (1987) Establishment of hybridoma secreting human monoclonal antibody against hepatitis B virus surface antigen. Biochem. Biophys. Res. Commun., 142, 805-812.
16. Levy, J.A., Virollainen, M., and Defendi, V. (1968) Human lymphoblastoid lines from lymph node and spleen. Cancer, 22, 517-524.
17. Kitano, K., Ichimori, Y., Sawada, H., Iwasa, S., Sasai, S., and Tsukamoto, K. (1991) Effective production of anti-tetanus toxid and anti-HBsAg human monoclonal antiboidies by serum-free culture of hybridomas. Cytotechnology, 5, S53-74.
18. Onodera, K., Etoh, Y., Takano, S., Morioka, H., and Shibai, H. (1987) Structure and function of E1a genes of adenovirus type 12. In: Umeda M et. al., (eds.) Biotechnology of mammalian cells, Jpn. Sci. Soc. Press, Tokyo and Springer Verlag, Berlin.
19. Iwamoto, K., Shintani, Y., Sasada, R., Honjo, T., and Kitano, K. (1990) Improvement in the antibody productivity of humna-human hybridomas by transfection with Tac gene. Cytotechnology, 4, 29-37.
20. Iwamoto, K., Shintani, Y., Sawada, H., and Kitano, K. (1991) Improvement in the proliferative activity of human-human hybridomas at low cell density by transfection with bFGF gene. Cytotechnology, 6, 90-103.
21. Kitano, K., Iwamoto, K., Shintani, Y., and Akiyama, S. (1988) Effective production of a human monoclonal antibody against tetanus toxoid by selection of high productivity clones of a heterohybridoma. J. Immunol. Methods, 109, 9-16.
22. Sato, G.H. (1975) The role of serum in cell culture. In: Litwick, G. 9eds.) Biochemical action of hormones. Vol. 3, pp391-396, Academic Press, New York.
23. Barnes, D.W., Sirbasku, D.A., and Sato, G.H. (1984) Cell culture methods for molecular and cell biology. Vol. 1-4, Alan R. Liss, Inc., New York.
24. Shintani, Y., Iwamoto, K., and Kitano, K. (1988) Polyethylene glycols for promoting the growth of mammalian cells. Appl. Microbiol. Biotechnol., 27, 533-537.
25. Murakami, H., Matsui, H., Sato, G.H., Sueoka, N., Chow, T.P., Kano-Sueoka, T. (1982) Growth of hybridoma cells in serum-free medium: Ethanolamine is an essential component. Proc. Natl. Acad. Sci., U.S.A., 79, 1158-1162.
26. Kitano, K., Shintani, Y., Ichimori, Y., Tsukamoto, K., Sasai, S., and Kida, M. (1986). Production of human monoclonal antibodies by heterohybridomas. Appl. Microbiol. Biotechnol 24, 282-286.
27. Takazawa, Y., Tokashiki, M., Hamamoto, K., and Murakami, H. (1988) High cell density perfusion culture of hybridoma cells recycling high molecular weight components. Cytotechnology, 1, 171-175.
28. Shintani, Y., Kohno, Y., Sawada, H., and Kitano, K. (1991) Comparison of culture methods for human-human hybridomas secreting anti-HBsAg human monoclonal antibodies. Cytotechnology, 6, 197-208.

RECEIVED May 17, 1995

Chapter 3

Antibody Production in Chinese Hamster Ovary Cells Using an Impaired Selectable Marker

R. S. Barnett, K. L. Limoli, T. B. Huynh, E. A. Ople, and M. E. Reff

Department of Gene Expression, IDEC Pharmaceuticals Corporation, 11011 Torreyana Road, San Diego, CA 92121

The expression levels of transfected genes in mammalian cells are primarily determined by the cellular DNA at the site of integration. In this report we describe immunoglobulin expression vectors designed to target mammalian loci that support high levels of expression. These vectors encode immunoglobulin light and heavy chain genes, the dihydrofolate reductase (DHFR) gene, and the dominant selectable marker neomycin phosphotransferase (NEO) gene. Selectivity of the vectors has been achieved by intentional impairment of the NEO gene by two different mechanisms. The NEO translation initiation site has been impaired and an intron has been introduced into the gene. Selection in G418 for clones containing the NEO gene yields an increased percentage of high expression clones from a decreased number of total clones. The majority of high expressing clones contain a single gene copy. Subsequent gene amplification results in clones producing very high levels of immunoglobulin with only a few gene copies.

It is our goal at IDEC to rapidly and efficiently generate stable transfected mammalian cells which produce high levels of immunoglobulin protein. Traditionally, immunoglobulin genes are transfected into recipient cells followed by selection for cells that have randomly integrated the genes into the cellular DNA. Levels of gene expression from cells following random integration of a gene on a plasmid are influenced by the effects of the local genetic environment at the site of chromosomal integration. This phenomenon is referred to as position effects (*1-6*). For immunoglobulins, the range of initial expression from clones derived from the same vector is from <25 ng/ml to >5,000 ng/ml (ME Reff, unpublished data).

Previously, our laboratory constructed an expression vector (TCAE) which was designed for the tandem expression of immunoglobulin light and heavy chain genes (*7*). This vector permits the simple insertion of immunoglobulin variable region DNA to produce chimeric (mouse-human or primate-human), humanized (CDR grafted), or all human antibodies (Figure 1). Figure 2 is a histogram showing expression results from individual clones in an experiment where TCAE encoding

0097–6156/95/0604–0027$12.00/0

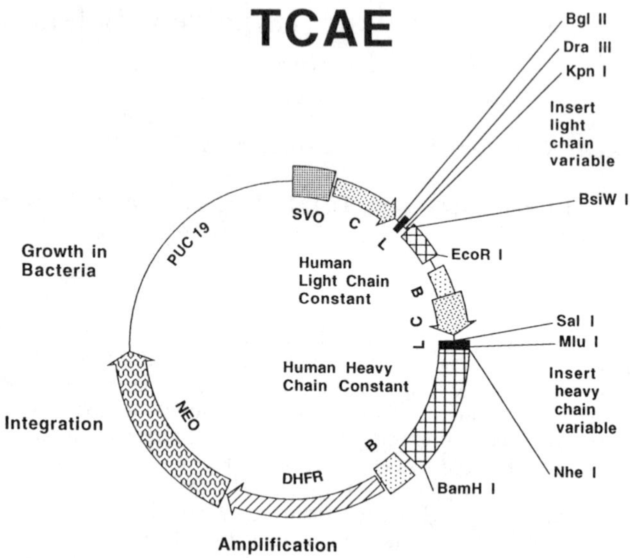

SVO = SV40 origin of replication
C = Cytomegalovirus promoter
L = Leader
B = Bovine growth hormone polyadenylation
DHFR = Dihydrofolate reductase
NEO = Neomycin phosphotransferase

Figure 1 Shown is an example of a modular expression vector used to produce immunoglobulin in mammalian cells. All of the components for integration, amplification and human constant regions of the heavy and light chain of immunoglobulin are on a single piece of DNA. We have versions with either human kappa constant or human lambda constant light chains, and human gamma 1 constant as well as human gamma 4 constant heavy chains.

Variable domains are inserted by generating polymerase chain reaction (PCR) fragments, and cloning into unique sites in the vector using either their own or synthetic leader sequences for secretion. All human, chimeric (mouse/human) and primatizied (monkey/human) antibodies have been created with these vectors. Vectors similar to this have been shown to work in a variety of cell types for both stable integration and amplification.

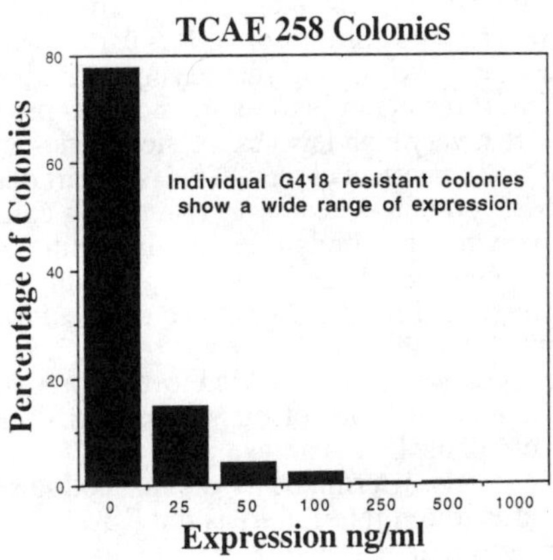

Figure 2 Shown is a histogram which plots the percentage of clones in 96 wells that expressed a given level of chimeric anti-CD20 antibody assayed as described in the text.

chimeric mouse-human anti-CD20 antibody was transfected into Chinese Hamster Ovary (CHO) cells. Two hundred fifty eight (258) individual G418 resistant colonies were assayed for immunoglobulin production in 96 well dishes. Colonies were assayed for expression by taking supernatant samples from confluent wells. Due to variations in cell number when assayed, antibody present in the well from residual media, and the experimental error of the ELISA assay itself, the absolute expression values obtained are not precise. However, the wide range of immunoglobulin expression level observed in this experiment, from less than 25ng/ml to greater than 500 ng/ml, can not be explained by solely by sampling variables. When expanded and more accurate expression values are obtained, clones still show large differences in expression levels.

Southern Blot analysis of five very high expression clones from five different experiments showed that three of the clones are single copy integrants (Figure 3). This implies that high immunoglobulin expression is not necessarily due to increases in the number of copies of plasmid integrated into the cell. Rather, the wide range of expression levels is attributable to the dominant effects of the chromosomal position on the randomly integrated plasmid DNA.

In this report we will refer to chromosomal loci that support very high levels of integrated gene expression as 'hot spots'. Identifying individual mammalian clones where a single copy insert results in high expression has proven to be an arduous task. Our data suggest that very high level expression clones are an extremely small proportion of the total G418 resistant clones. This is in part due to the fact that only a minute amount of NEO protein is necessary to enable a cell to survive selection. Thus, cells can survive when plasmids integrate in positions where there is low expression of the NEO gene and adjacent immunoglobulin genes. Positions where there is very high expression of the NEO gene and adjacent immunoglobulin genes are rare within the mammalian genome. As a result, it is necessary to screen hundreds to thousands of clones to identify high level expression isolates.

Here we report the modification of our expression vectors by intentionally impairing the neomycin phosphotransferase gene used for selection. Eukaryotic translation initiation is largely determined by the nucleotide sequence surrounding the translation start codon, often referred to as the Kozak sequence (*8*). We have mutated the NEO gene to create a 'bad' Kozak sequence in an effort to reduce productive translation of the NEO messenger RNA. Independently, or in addition, we have introduced an artificial intron within the NEO coding region in another attempt to impair selection by NEO.

The effects of creating an impaired dominant selectable marker are twofold. Since most single copy integrants will not express enough NEO to survive selection, the overall number of G418 resistant cells is greatly reduced, facilitating screening. A higher percentage of the clones surviving selection are those in which the impaired NEO gene has been integrated into 'hot spots' within the genome, which concomitantly yield very high levels of linked gene expression.

Once isolated, transfectants which display a very high level of immunoglobulin protein production are induced to undergo gene amplification by selection in Methotrexate (MTX) for the dihydrofolate reductase (DHFR) gene (*9*). As the DHFR gene copy number increases through amplification, there is a parallel increase in the closely linked immunoglobulin gene copy number with an accompanying rise in immunoglobulin production. Amplification of initially very high level expression clones yields cells producing even greater levels of immunoglobulin protein from a minimal number of gene copies.

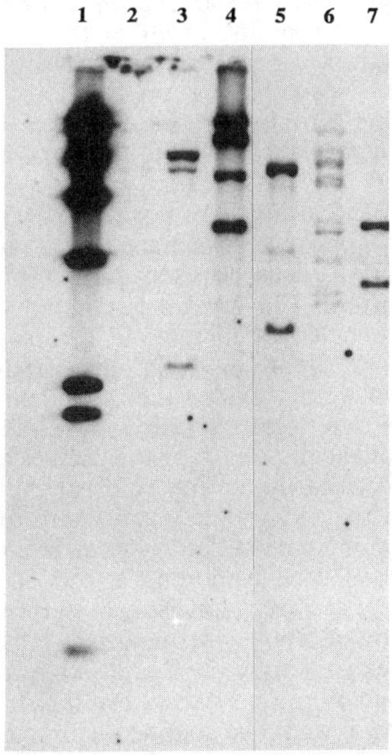

Figure 3 Shown is a Southern Blot of DNA isolated from the highest expressing G418 clones from five different electroporations into CHO with five different plasmids. 5 mg of high molecular weight DNA digested with EcoR I, which cuts once in each plasmid, gives two bands of different sizes for each integration site into cellular DNA.

Lane	DNA	Plasmid Copies	NEO gene
1	Molecular weight markers		
2	CHO parent		
3	TCAE clone	single	consensus Kozak
4	ANEX clone	two	translation impaired
5	GKNEOSPLA clone	single	splice
6	NEOSPLA clone	five	translation impaired and splice
7	NEOSPLA clone	single	translation impaired and splice

ANEX: Vectors with a Impaired Translation Initiation Site

Following is a description of the translational impairment of the NEO gene. The consensus translation initiation sequence for eukaryotic genes is (-3)AccATG (+1)G where both the (-3) and the (+1) positions, relative to the ATG start codon for methionine, are most often a purine nucleotide (*8*). While it is not uncommon for a eukaryotic gene to have a pyrimidine in either the -3 or the +1 position, it is extremely rare for a eukaryotic gene to have a pyrimidine nucleotide in both positions (*8*). A eukaryotic gene having a pyrimidine in both positions will be referred to as a fully impaired Kozak sequence.

The neomycin phosphotransferase (NEO) gene of the transposon TN5 is a prokaryotic gene, and its translation start site is (-3)CgcATG(+1)A, i.e. a pyrimidine in position -3 and a purine in position +1. The translation initiation sequence for the neomycin phosphotransferase (NEO) gene of the TCAE vector was engineered to be the Kozak consensus sequence (-3)AccATG (+1)G, changing the -3 pyrimidine to a purine. The ANEX expression vector was derived from TCAE by employing oligonucleotide directed PCR mutagenesis to alter the NEO gene Kozak sequence to create a fully impaired consensus Kozak (-3)TccATG(+1)C. These changes also alter the second amino acid of the enzyme, from an isoleucine in TN5 (ATT), to a valine in TCAE (GTT), to a leucine in ANEX (CTT). These amino acid changes by themselves are not expected to affect NEO since others have altered the amino acids at the amino terminus of the neomycin phosphotransferase gene without impairing its function (*10*). In addition, in ANEX, an upstream out-of-frame initiation codon in a perfect consensus Kozak has been introduced. Translation beginning at this start codon should impair translation from initiating at the correct downstream start codon (*11*).

Furthermore, this region of DNA containing the translation start codon in the ANEX vector has been designed to allow the possible formation of a secondary 'stem-loop' or 'hairpin' structure in the transcribed RNA. In this potential RNA structure, the NEO start codon would be confined within the complementary region forming the stem of the hairpin structure, while the upstream out-of-frame start codon would be in an accessible loop. We have, however, not done experiments to confirm that this RNA structure exists.

The ANEX vector containing the mouse-human chimeric anti-CD20 antibody was electroporated into CHO cells (9 electroporations, 25 mg/4.0 x 10^6 cells/electroporation) and the cells were selected for G418 resistance in 96 well dishes. Expression of antibody from the NEO resistant clones obtained in this experiment was compared to the earlier data from the TCAE vector experiment (Figure 4). These were independent experiments carried out at separate times, therefore, comparing expression levels offers only an approximate comparison. A dramatic increase in expression of immunoglobulin from individual ANEX colonies was seen when compared with TCAE clones. The translation impaired ANEX vector yielded 121 G418 resistant colonies. Twenty (20) of the colonies (16.5%) expressed greater then 1000 ng/ml of antibody. Only 5 of the 121 colonies (4.1%) expressed less than 25 ng/ml of antibody. In contrast, two electroporations into CHO (25 mg/4.0 x 10^6 cells/electroporation) of the TCAE vector, which contains a consensus Kozak sequence at the translation initiation site, yielded 258 G418 resistant colonies. There were no colonies in this experiment that produced greater than 1000 ng/ml of antibody. Two hundred and one (201) of these colonies (78%) expressed less than 25 ng/ml of antibody.

In other experiments where equal amounts of DNA of the two vectors (TCAE, ANEX) were transfected simultaneously, the ANEX vector consistently yielded a 80-90% reduction in numbers of G418 resistant clones per electroporation.

Therefore, by utilizing a translationally impaired dominant selectable marker, the number of colonies to be screened decreased, while the amount of linked immunoglobulin gene product significantly increased.

NEOSPLA: Artificial Intervening Sequence within NEO

It is well known that the majority of prokaryotic genes typically possess neither intervening sequences (introns) nor undergo RNA splicing. In an additional effort to impair the NEO gene used for dominant selection, an artificial intron was introduced into the coding region of NEO, and the immunoglobulin and DHFR genes were inserted into the intron. It was postulated that the sequences coding for RNA cleavage and polyadenylation contained within the light chain, heavy chain and DHFR genes might lead to truncation of the nascent NEO message before splicing could occur. Such cleavage of the NEO pre-mRNA would result in fewer mature NEO transcripts and decreased levels of functional NEO protein.

A 47 base synthetic oligonucleotide linker (Figure 5) was inserted within the neomycin gene coding sequence of both TCAE and ANEX. Following insertion, the linker and surrounding sequences code for consensus splice sequences including the 5' splice donor, branch point, polypyrimidine tract and 3' splice acceptor sites (*12,13*). In addition, a unique Not I restriction site was placed between the 5' donor site and the branch point. Two new vectors were then generated by insertion of a light chain immunoglobulin gene, a heavy chain immunoglobulin gene and a DHFR gene into the Not I site in the same transcriptional orientation with respect to NEO. The vector derived from TCAE containing a consensus Kozak sequence for the NEO translation initiation codon is referred to as 'GKNEOSPLA'. The vector containing a fully impaired Kozak and an upstream out-of-frame start codon (ANEX like), is referred to as 'NEOSPLA'. Figure 6 is a schematic representation of the NEOSPLA vector.

Twenty five (25) mg of each plasmid (TCAE, GKNEOSPLA, NEOSPLA) encoding the chimeric mouse-human anti-CD20 antibody was transfected via electroporation into 4 x 10^6 CHO cells and plated into 96 well tissue culture plates. Based upon the expected frequency of G418 resistant colonies from preliminary experiments, 400,000 cells were plated from the TCAE vector, 2,000,000 cells were plated for the GKNEOSPLA vector, and for NEOSPLA, the entire electroporation of 4,000,000 cells was plated. An identical number of G418 resistant colonies (16) was obtained from all three vectors. The effects on expression were dramatic (Figure 7). One of the GKNEOSPLA clones was above the linear range of our ELISA at a 1/20 dilution (>2000 ng/ml) and two of the NEOSPLA clones were similarly off scale.

Northern Blot analysis of the NEOSPLA clone probed with either the entire NEO gene or the second exon of the NEO gene shows only a single NEO message which is identical in size to the NEO message from an unspliced vector (Figure 8), indicating that the artificial splice is functional. Unspliced message was not seen in this experiment.

High level expression of other immunoglobulins and of other non-immunoglobulin single chain proteins has been demonstrated using these vectors as well as vectors with similarly impaired alternative selectable markers in our laboratory. (Data not shown)

Amplification of a High Expression Cell Line Containing a Single Integrated Plasmid

A NEOSPLA vector containing a chimeric primate-human antibody was transfected into CHO. Seventy three (73) colonies arising from five transfections

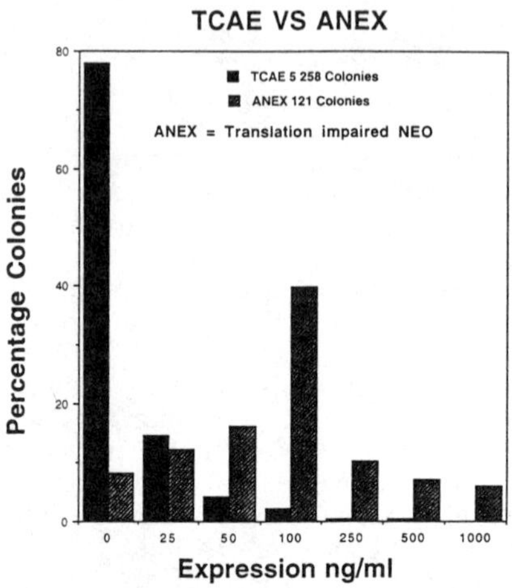

Figure 4 Shown is a histogram which plots the percentage of clones in 96 wells that expressed a given level of chimeric anti-CD20 antibody assayed as described in the text.

```
                        Synthetic Splice

5'  CAG GTAAGT GCGGCCGC TACTAAC TCTCTCCTCCCTCCTTTTTCCT GCAG G  3'
  5' Splice Donor        Branch Point  Polypyrimidine Tract   3'splice Acceptor
              Not I
              Cloning
              Site
```

Figure 5 Shown is the nucleotide sequence encoding an artificial splice with a site (Not I) for the insertion of additional DNA.

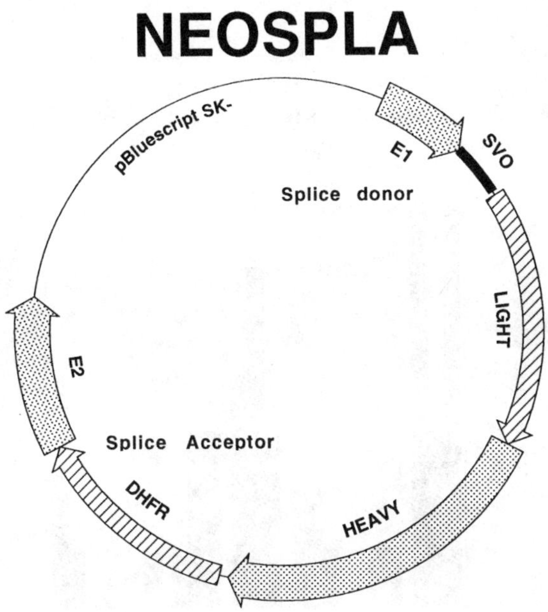

E1 = Neomycin phosphotransferase Exon1
E2 = Neomycin phosphotransferase Exon2
SV0 = SV40 Origin
Light = Immunoglobulin light chain
Heavy = Immunoglobulin heavy chain
DHFR = Dihydrofolate reductase

Figure 6 Schematic representation of the NEOSPLA vector.

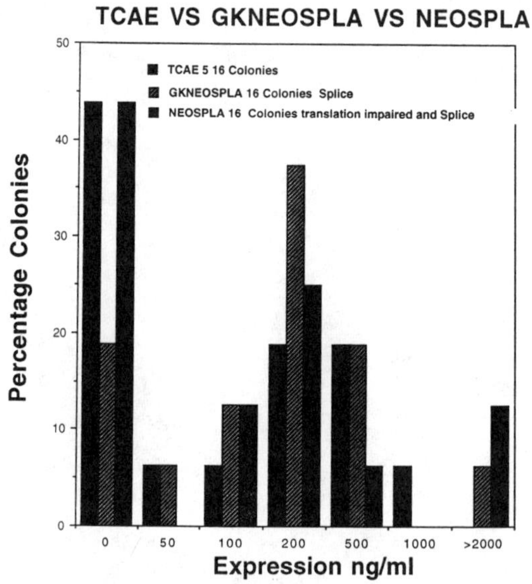

Figure 7 Shown is a histogram which plots the percentage of clones in 96 wells that expressed a given level of chimeric anti-CD20 antibody assayed as described in the text.

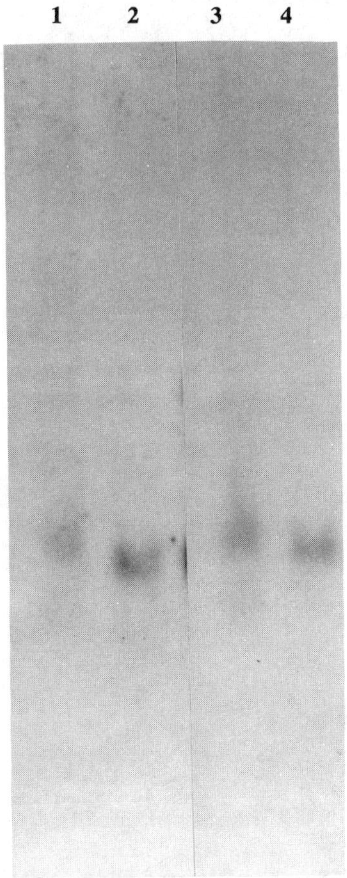

Figure 8 Shown is a Northern blot of RNA isolated from a G418 resistant TCAE clone and NEOSPLA clone. Both clones secreted high amounts of chimeric antibody.

	RNAs	
Lanes 1, 3	TCAE vector	unspliced NEO
Lanes 2, 4	NEOSPLA vector	spliced NEO

	Probes
Lanes 1, 2	Second Exon of NEO
Lanes 3,4,	Entire NEO

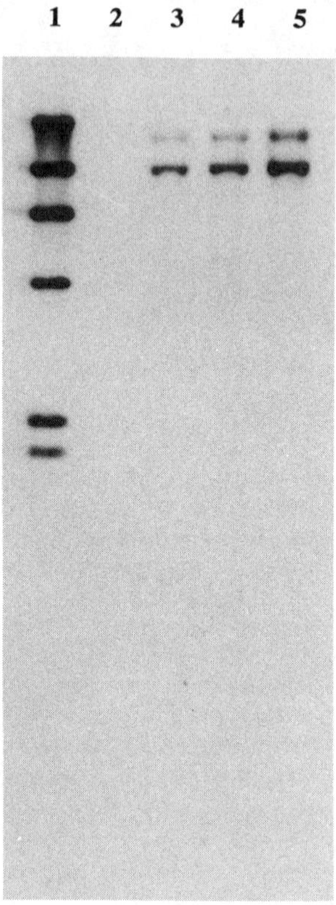

Figure 9 Shown is a Southern Blot of DNA isolated from a CHO clone integrated into a 'hot spot' and the derived 5nM and 50nM MTX clones. 5 mg of high molecular weight DNA was digested with Nhe I which cuts once in each plasmid and gives two bands of unknown size for each integration site into cellular DNA.

Lane	DNA	Expression Level
1	Molecular weight markers	
2	CHO parent DG44	
3	G418 clone	5 mg/liter/spinner
4	5 nM MTX clone	30 mg/liter/spinner
5	50 nM MTX clone	150 mg/liter/spinner

were screened, and the highest immunoglobulin expresser was expanded. This clone was shown by Southern Blot to be a single copy integrant (Figure 9) and it was secreting approximately 2.2pg of immunoglobulin/cell/day.

Amplification of this clone in 5nM MTX was performed and a resulting clone producing 16pg/cell/day was isolated. This 5nM clone was then amplified in 50nM MTX, and a clone producing 45 pg/cell/day was isolated. Southern Blot analysis (Figure 9) shows the amplified DNA of the 5nM and 50nM clones is in the same integration site as the unamplified G418 parent, and that the number of copies of the plasmid in the 50nM clone is very low. The elapsed time from the introduction of the plasmid into the CHO cell until the expanded 50 nM clone was frozen for storage was a relatively short seven months.

In seven day antibody production runs in spinner culture in serum free media, where cells were seeded at 3×10^5 viable cells/ml and peak cell densities were less than 2×10^6 cells/ml, this 50nM clone produced over 150 mg/liter both with and without MTX. Stability of expression in the absence of MTX is a characteristic of these cell lines selected in low levels of MTX and containing a low plasmid copy number. In addition, other 'hot spot' clones producing similar amounts of antibody in spinner cultures have been shown to produce over a kilogram of antibody in a six day fermenter run in a 2500 liter fermenter at low cell densities.

Discussion

The most rapid method to create a mammalian clone expressing high levels of antibody is to insure that the plasmid DNA containing the immunoglobulin genes is integrated in a location within cellular DNA where transcription is maximal. While these clones do not secrete as much antibody per cell as amplified cell lines, they can be quickly amplified with low levels of methotrexate to reach maximal levels of secretion, which for immunoglobulins are about 50 to 100 pg of antibody/ cell /day (*7,14-16*).

Large regions of the mammalian genome are organized into heterochromatin, which is believed to be transcriptionally inactive (2), and a plasmid that integrates into these regions will probably not make enough NEO to survive selection. Our data suggest that the chances of a plasmid locating a 'hot spot' in transcriptionally active DNA are between 1/100 and 1/1000 of the total selected random integrations. By impairing the neomycin phosphotransferase gene, we have constructed plasmids that reduce the number of G418 resistant clones by about 50 fold. Approximately 5 to 10 per cent of these G418 resistant clones are single copy integrants in 'hot spots'. CHO clones which have been derived from a 'hot spot' integration site and then minimally amplified have produced 800 mg/liter of antibody in a 100 liter fermenter in six days at low cell density (2 to 3×10^6 cells/ml).

These strategies for rapidly achieving high level production of antibody are useful for other proteins that need to be produced at high levels in mammalian cells.

We are in the process of marking 'hot spots' in CHO cells, and returning to them via homologous recombination to ensure DNA integration gives single copy high expression.

Literature Cited

(1) Grosveld, F.; van Assendelft, G. B.; Greaves, D. R.; Kollias, G. *Cell* **1987**, *51*, 975-985.

(2) Wilson, C.; Bellen, H. J.; Gehring, W. J. *Annu Rev Cell Biol* **1990**, *6*, 679-714.

(3) Conklin, K. F.; Groudine, M. *Molecular and Cellular Biology* **1986**, *6*, 3999-4007.
(4) von Knebel Doeberitz, M.; Bauknecht, T.; Bartsch, D.; zur Hausen, H. *Proc Natl Acad Sci* **1991**, *88*, 1411-1415.
(5) Al-Shawi, R.; Kinnaird, J.; Burke, J.; Bishop, J. O. *Molecular and Cellular Biology* **1990**, *10*, 1192-1198.
(6) Yoshimura, F. K.; Chaffin, K. *Molecular and Cellular Biology* **1987**, *7*, 1296-1299.
(7) Reff, M. E.; Carner, K.; Chambers, K. S.; Chinn, P. C.; Leonard, J. E.; Raab, R.; Newman, R. A.; Hanna, N.; Anderson, D. R. *Blood* **1994**, *83*, 435-445.
(8) Kozak, M. *Nucleic Acids Research* **1987**, *15*, 8125-8132.
(9) Kaufman, R. J.; Sharp, P. A. *J. Molecular Biology* **1982**, *159*, 601-621.
(10) Thomas, K.; Capecchi, M. R. *Cell* **1987**, *51*, 503-512.
(11) Kozak, M. *Molecular and Cellular Biology* **1987**, *7*, 3438-3445.
(12) Mount, S. M. *Nucleic Acids Research* **1982**, *10*, 459-472.
(13) Zhuang, Y.; Goldstein, A. M.; Weiner, A. M. *Proc Natl Acad Sci* **1989**, *86*, 2752-2756.
(14) Shitara, K.; Nakamura, K.; Tokutake-Tanaka, Y.; Fukushima, M.; Hanai, N. *Journal of Immunological Methods* **1994**, *167*, 271-278.
(15) Page, M. J.; Sydenham, M. A. *Bio/Technology* **1991**, *9*, 64-68.
(16) Fouser, L. A.; Swanberg, S. L.; Lin, B.-Y.; Benedict, M.; Kelleher, K.; Cumming, D. A.; Riedel, G. E. *Bio/Technology* **1992**, *10*, 1121-1127.

RECEIVED May 17, 1995

Chapter 4

Antibody Production in Insect Cells

Kathleen N. Potter, Yucheng Li, and J. Donald Capra

Department of Microbiology, University of Texas Southwestern Medical Center, Dallas, TX 75235

The baculovirus expression system produces complete, biologically active human, mouse, human-mouse hybrid and human-mouse chimeric antibodies of the IgG, IgM and IgA isotypes in insect cells. Antibodies of all three isotypes are secreted as four-chain monomers. IgA antibodies can form dimers to a limited extent when coexpressed with J chain, while IgM antibodies predominantly remain as four-chain monomers, even in the presence of J chain. The tools of molecular biology allow the insect system to be manipulated to produce natural antibodies as well as those with point-mutations, exchanged domains and regions, and unnatural heavy and light chain pairings suited for immunoglobulin structure/function studies, diagnosis and therapy. The time between the initial transfection with the transfer vector and antibody expression is approximately 3 weeks.

Immunoglobulins are four-chain glycoproteins composed of two identical heavy (H) and two identical light (L) chains (reviewed in *1*). The H chains are linked through interchain disulfide bridges and each L chain is disulfide bonded to one H chain to form a H_2L_2 monomer (Figure 1). Each H and L chain can be subdivided into a variable (V) region located at the amino-terminal ends and a constant (C) region at the carboxy-terminal ends. The V regions are further subdivided into areas of relatively conserved sequences called framework regions (FR) and stretches which are more hypervariable in sequence referred to as complementarity-determining regions (CDR). In both the H and L chains, the V regions have the alternating pattern of FR1,CDR1, FR2, CDR2, FR3, CDR3 and FR4. The non-covalent association of the six CDRs from the H and L chain V regions forms the antigen combining site and determines the specificity of the antibody. Each monomeric antibody has two antigen combining sites, one located at the tip of each heavy-light chain pair. A dimer consisting of only the V_H-V_L regions is referred to as Fv, also has antigen binding capability.

0097–6156/95/0604–0041$12.00/0

As immunoglobulins are proteins, they are themselves antigenic. Antigenic determinants located in the V region are referred to as idiotypes (Id). Idiotypic determinants can induce anti-idiotypic antibodies within the same or different host, and have been postulated to comprise a self-regulatory immune mechanism operating through a network of interacting immunoglobulins (*2*). Antigenic determinants located in the C region are called isotypes, and are used to group immunoglobulin H and L chains into isotypes or classes. There are five human H chain isotypes; mu (μ), alpha (α), delta (δ) and epsilon (ε), and two light chain isotypes called kappa (κ) and lambda (λ). In nature, secreted IgG antibodies are monomeric while IgM antibodies are found predominantly as pentamers and hexamers. IgA antibodies are found predominantly as monomers in serum but mostly as dimers or tetramers in secretions (*1*).

The basic structural unit of immunoglobulins is the immunoglobulin domain. The L chain is composed of two domains, V_L and C_L, and the IgG H chain has four domains, V_H, C_H1, C_H2 and C_H3. IgM and IgE heavy chains have a fifth domain C_H4. Each domain contains approximately 110 amino acid residues. A dimer of the C_H2-C_H3 domains bound by disulfide bonds is called the fraction crystallizable or Fc fragment. The immunoglobulin domain structure can be manipulated by exchanging whole domains or portions of domains between antibodies.

Synthetic antibodies have been produced in many expression systems including bacteria, yeast, plant, mammalian and insect cells (reviewed in *3*). Some of the difficulties involved in studying human antibodies are the result of technical problems such as obtaining stable human-human hybridomas and the lack of secreted monoclonal antibodies in amounts sufficient for analysis. Monoclonal antibodies with defined specificity are very useful diagnostic reagents, however, they are not widely used in human therapy due to the fact that they are most often produced in mice and are immunogenic when injected into humans. A further limitation to the use of monoclonal antibodies for some immunoglobulin analyses is that the antibodies produced are limited to those which the immune system can generate naturally. This precludes the production of human-mouse chimeric antibodies, the production of many combinatorial antibodies using non-parental H and L chain pairings, and desired FR and CDR unions. Over the past few years the sequences of a large number of immunoglobulins from a variety of species have been determined (*4*). Significant advances have been made in the area of humanizing antibodies in which the CDRs are derived from a non-human (usually murine) origin while the FR and C regions are human in sequence. In general, humanized immunoglobulins have reduced immunogenicity, but they do not have as high a binding affinity to target antigens as do the parental non-human antibodies.

The baculovirus system is capable of producing full-length H and L immunoglobulin chains, secreting complete four-chain heterodimeric monomers (*5,6*) as well as single-chain Fv fragments (*7,8*). The ability of eukaryotic insect cells to produce full-length C regions permits in addition to the study of antigen combining sites, the analysis of the effector functions associated with Fc regions such as complement fixation and Fc receptor binding. These *in vitro* engineered antibodies can be produced in quantity sufficient for X-ray crystallographic analysis, detailed structure/function analyses, diagnosis and human therapy (*9*).

We will begin this chapter with a short over-view of the baculovirus system itself, followed by a description of how the system is being used for the production and analyses of antibodies.

The Baculovirus Expression System

Insect Viruses. The baculovirus expression system uses an insect virus genome to carry foreign genes into insect cells which are transcribed and translated along with the normal viral genes. The insect virus most commonly used as vector is the multiply embedded nuclear polyhedrosis virus *Autographa californica* (AcMNPV). AcMNPV belongs to the family *Baculoviridae*, a group of double-stranded, circular, supercoiled DNA viruses which replicate in the nuclei of permissive insect cells. In nature, the infection process begins when an insect ingests polyhedral crystals containing infectious virions. The crystals dissolve in the alkaline midgut of the insect releasing the occluded viruses (OV). The occluded viruses consist of several nucleocapsids within a single envelope, hence the term multiply embedded viruses. The enveloped viruses proceed to enter the midgut cells where the infection process begins. The time course of the viral infectious cycle has been determined in tissue culture using AcMNPV in the Sf21 cell line. DNA replication starts at approximately 6 hr post-infection and by 10-12 hr post-infection, single nucleocapsids bud from the plasma membrane producing infectious, enveloped virions. These extracellular viruses (EV) are responsible for the spread of infection within an insect. From 20-72 hr post-infection, enveloped nucleocapsids retained in the cell nucleus are encased within polyhedral crystalline matrices called occlusion bodies (OB). Occlusion bodies are composed primarily of a single structural protein called polyhedrin, which has a molecular weight of 29 kDa. Polyhedrin is expressed at very high levels within the cells accounting for 25-50% of the total cellular protein. The OB are responsible for viral spread between insects under natural conditions of virus transmission. When the body of an infected insect decomposes, the OB weather environmental forces until they are ingested by healthy insects and the infection cycle begins again. In tissue culture, both EV and OB are produced, but it is the EV which are responsible for the cell-to-cell spread of infection within a single flask and all subsequent infections of *in vitro* cultured cells.

Both polyhedrin, and the p10 encoded protein comprising fibrillar structures, are non-essential for viral infection and replication (*10*). The genes encoding these proteins can, therefore, be replaced by numerous foreign genes under control of their very strong promoters. Since the viral genome is large consisting of approximately 128-134 kb, it is not possible to clone directly into the viral DNA. Instead a two-step process has been developed (described in detail in *11,12,13*) which involves cloning the gene of interest into a transfer vector which is amplified in *E. coli* and then transfected into insect cells in the presence of wild-type viral DNA.The transfer vector contains regions of viral sequence flanking the foreign gene, and by intracellular homologous recombination, the foreign gene is transferred into the viral genome at a site determined by the flanking viral sequences in the transfer vector. When the flanking sequences are polyhedrin gene derived, the foreign gene replaces the polyhedrin gene and the net result is that the recombinant viruses replicate and produce EV but not OB. As OB are readily seen under the light microscope, it is possible to distinguish between wild-type virus, occlusion-plus infected cells and recombinant virus, occlusion-minus infected cells.The efficiency of recombination is improved when linearized wild-type DNA is used instead of circular DNA (*14*). Depending on the linearized wild-type DNA used, the recombination efficiency can range from 30% to nearly 100% compared with the 0.01-0.001% with circular viral DNA.

Purified viral stocks are generated by infecting cells at a multiplicity of infection (MOI) of 0.01- 0.1 and harvesting the supernatant at three days post-infection. We routinely use virus at passage three for antibody production. We do not passage the virus at high MOI or beyond the fifth passage as it has been demonstrated that viral mutants which produce few polyhedra (FP) are selected in tissue culture (15) resulting from the insertion of lepidopteran transposons into the fp25 locus (16,17,18). This FP type of plaque can be confused with recombinant virus plaques. Baculoviruses also produce defective interfering particles after serial undiluted passage in cell culture (19) and continuous production in bioreactors (20) which has the effect of reducing infectious viral titers. If necessary, it is possible to replaque a viral stock to select for infectious, wild-type or recombinant virus. Viral stocks are very stable and can be stored for several years at 4°C without a significant drop in titer. One study reports that the virus is not appreciably affected by multiple freeze-thaw cycles, but is more sensitive to inactivation by light (21). These viruses can be quantitated either by plaque assay using an agarose overlay or by the endpoint method (22). A dye, MTT (3-[4,5-dimethylthiazol-2-y1]-2,5-di phenyl- tetrazolium bromide), has been used to stain background cells so that plaques can be better seen and more accurately counted for quantitation (23).

Insect Host Cells. The three most commonly used lepidopteran cell lines for the expression of foreign proteins are derived from ovarian cells from the fall armyworm, *Spodoptera frugiperda* and the cabbage looper, *Trichoplusia ni*. These are the Sf21 IPBL-SF-21 AE line, and a clonal derivative Sf9 (24), TN-368 (25) and BTI-TN-5B1-4 cells (High-Five^{TM} cells; Invitrogen, San Diego, CA) which are derived from *T. ni* cells.

The *T. ni* and *S. frugiperda* cell lines can be grown both as attached monolayers and suspension cultures, both in media containing 10% fetal calf serum as well as serum-free media of various compositions. Cell densities and viabilities vary depending on the medium composition and the vessel used for suspension cultures (26). Usually cells in suspension are grown to a density of 1-3 x 10^6 cells/ml, however, there are reports of Sf9 being grown to more than 5 x 10^6 cells/ml with greater than 95% viability in a spinner vessel (27,28). Recently several companies have reported that Sf9, Sf21 and BTI-TN-5B1-4 cells can be grown to more than 8 x 10^6 cells/ml in shake-flask cultures in the presence of Pluronic F-68. The level of dissolved oxygen is a critical component in obtaining high cell density with high viability. The ability to grow cells viably to high density potentially allows for a more cost-efficient production of foreign proteins, provided that all of the cells maintain the same output of protein as at the lower cell densities.

Invertebrate cells are capable of many post-translational modifications to expressed foreign proteins. These include removal of signal peptides, proper trafficing through the cell, glycosylation, disulfide-bridge formation and secretion. All of these complex cellular functions are critical to the production of immunoglobulins. Protein sequencing determined that signal peptides are removed from secreted immunoglobulin chains (4). Carbohydrate analysis of human plasminogen produced in Sf21 cells was the first demonstration that trimming and processing of high-mannose carbohydrate into complex-type oligosaccharides can occur in these cells (29). It is now established that insect cells can add both N- and O-linked carbohydrates (30).The functional expression of human plasminogen is also a good example of the ability of Sf21 cells to form disulfide bonds, as the protein contains 23 disulfide bridges.

AcMNPV infected cells in culture demonstrate a pronounced cytopathological effect (CPE). This is characterized by the rounding of the spindle-shaped *T. ni* cells, enlarging of the cells, an increase in nuclear size and a noticable reduction in the cytoplasmic area. It is perhaps more difficult to observe the CPE in Sf9 and Sf21 cells, as these uninfected cells are round to begin with. The production of polyhedra, however, is very obvious under the light microscope in all three cell lines as the presence of approximately 70 OB per cell gives the nucleus a black appearance.

Virus production and the post-translational modification of proteins in cultured insect cells are influenced not only by the cells themselves but by the growth medium used. We have not observed any difference in virus infectivity between our Sf9 cells adapted to serum containing medium versus those adapted to serum free medium. Certainly for the production of human immunoglobulins, their secretion into a medium lacking bovine antibodies simplifies the purification and quantitation processes. Although several hundred foreign proteins have been expressed in invertebrate cells, the production levels and functionality of expressed proteins are still determined on a case-by case basis.

Foreign Genes and Transfer Vectors. Insect cells are not capable of efficiently splicing out introns (*31*). It is necessary, therefore, to clone genes which either naturally do not contain introns or cDNA molecules. Transfer vectors for the baculovirus expression system are continually being made and many are commercially available. Critical features of an insect transfer vector are an *E. coli* origin of replication for amplification in bacterial cells, a drug resistance marker such as the β-lactamase gene encoding ampicillin resistance, a phage sequence such as f1or M13 for the production of single-stranded DNA and enough flanking viral sequences for homologous recombination into wild-type viral DNA. It is now possible to place foreign genes under the control of a choice of promoters which regulate the time in the virus infection cycle at which the foreign gene is transcribed and translated. Genes are placed either under control of late or very late promoters, depending on the post-translational modifications required to produce a functional protein, as the number of modifying enzymes present varies during the infection cycle. The most commonly used late promoters are the 39K and the basic protein promoters, while the frequently used very late promoters are the polyhedrin and p10 promoters. Some transfer vectors are designed for the coexpression of two, three and four different proteins (reviewed in *12* and *32*, *33*). There are vectors which allow for a color distinction between wild-type and recombinant virus plaques, thereby reducing the effort required to plaque purify recombinant viruses. Some vectors also contain sequences which result in the production of recombinant proteins fused to tags which are useful in the rapid purification of the recombinant protein from insect cellular and viral proteins. Some of the tags which have been used in the baculovirus system include the glutathione-S-transferase sequence (*34*) and the hexahistidine tag (*35*).

Transfection. Fundamental to the operation of this expression system is the fact that AcMNPV DNA is infectious to insect cells (*36,37,38*). The process of transfection usually involves the simultaneous introduction of the transfer vector containing the gene of interest and wild-type virus DNA into the same cell. Homologous recombination between the vector and the wild-type virus DNA results in the viral genome containing the foreign gene. There are several ways to introduce DNA into insect cells. Transfection of insect cells is often performed using calcium phosphate precipitation (*36,37,38*), and electroporation has been reported as an efficient method (*39*). There is also a report indicating that transfection of the transfer vector into cells pre-infected with wild-type virus using the calcium phosphate precipitation method can be used for producing recombinant baculoviruses (*40*). In our laboratory we use cationic liposomes and wild-type DNA linearized in the polyhedrin gene (Invitrogen, San Diego, CA) to place antibody genes under control of the polyhedrin promoter. Although it is often recommended to use cesium chloride density gradient or anion-exchange chromatography purified vector DNA for transfection, we routinely use DNA isolated directly from *E. coli* by phenol extraction and ethanol precipitation without difficulty. The efficiency of recombination obtained using the linear wild-type DNA is sufficiently high that occlusion-minus plaques are readily detected using a dissecting microscope, even in the absence of color production (*41*).

We determine the presence of the foreign gene in the virus by a direct PCR on the virus. Several microliters of clarified viral supernatant are first heated at 94°C for 5 minutes. 1 µl is used as template in a standard PCR reaction mix using *Taq* polymerase (Perkin-Elmer, Cetus, Norwalk, CT). The pre-heating step and the small amount of virus supernatant are the critical factors in making this reaction work. The PCR products are then analysed by agarose gel electrophoresis to determine the size of the insert. The PCR products are also good substrates for DNA sequencing.

Antibody Expression in the Baculovirus System

Cloning and Mutagenesis of Antibody Genes in the Baculovirus System. The first examples of intact antibody production in insect cells involved the synthesis of mouse IgG1/κ anti-arsonate antibodies (*5*) and mouse IgG2a/κ anti-*Pseudomonas aeroginosa* lipoprotein I (*6*). To date, complete antibodies of the IgM (*42*), IgG (*5,6*) and IgA (*43*) isotypes, as well as single-chain molecules (*7,8*), have been produced in the baculovirus system. We are currently using AcMNPV for the production of complete, four-chain human and mouse-human chimeric antibodies in Sf9 and BTI-TN-5B1-4 cells essentially as described (*44,45*), with some modifications in order to reduce the time between gene cloning and protein expression. The sources of human immunoglobulin genes in our laboratory are hybridoma cells producing antibodies of known sequence and specificity. A flow diagram of the system is shown in Figure 2. Total RNA is extracted from hybridoma cells and cDNA is produced using either random hexamer or oligo-dT primers. L chains, consisting of both V and C regions are PCR amplified using primers specific for the 5' and 3' sequences of the genes using either Taq (Perkin-Elmer, Cetus, Norwalk, CT) or Pfu (Stratagene, La Jolla, CA) DNA polymerase. The enzyme Pfu is preferred as it has a lower error frequency. The primers contain a 5' NcoI and a 3' XbaI restriction enzyme site respectively. Complete L chains can be amplified in a single PCR reaction. Heavy chains are constructed by amplifying the V_H portion in one reaction which is then attached by PCR-Splicing by Overlap Extension (PCR-SOEing) (*46*) to a C region obtained in a separate reaction. The PCR amplified genes are ethanol precipitated, digested with the requisite restriction enzymes, purified by electrophoresis through a low melting temperature agarose gel and ligated overnight into the transfer vector pAc360NX (*5*). Melted agarose containing the DNA bands is added directly to the ligation mixture. The ligation products are tranformed into *E. coli* (we generally use XL1-Blue, Stratagene, La Jolla, CA), and transformants are screened for the presence of the gene by plasmid mini-preparations and restriction enzyme digestion. Site-directed mutations are introduced into genes by placing base alterations in the priming oligomers (*47*). Exchanges between regions of two antibodies are accomplished by PCR amplification of the desired segments using overlapping oligomer sequences and producing conitguous sequences by PCR-SOEing. The PCR products are then digested with restriction enzymes, purified by electrophoresis and ligated into the transfer vector. Cells are transfected as described above following the manufacturer's instructions, and the supernatant is harvested and analysed by plaque assay. Occlusion minus plaques are picked and each agarose plug is dispensed into 1 ml of serum plus medium. An aliquot (200 µl) is used to infect 2 x 10^5 cells/well in 24 well plates. At 3-5 days post-infection, the cell pellets are analysed for protein synthesis by Western blot. The picked plaques are also

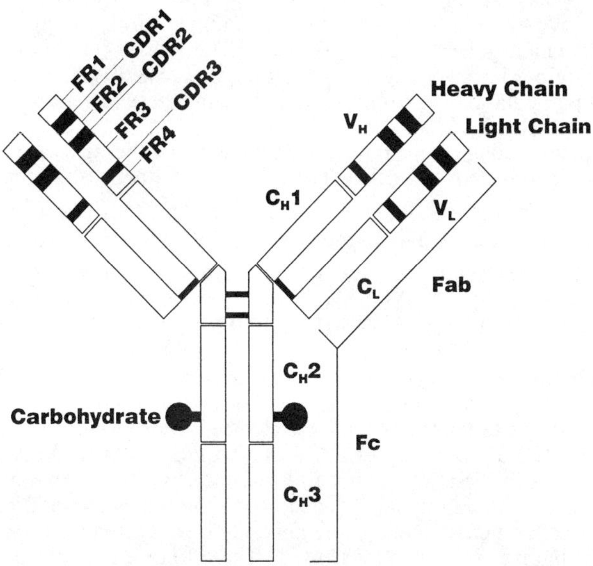

Figure 1. A representation of the polypeptide chain structure of a human IgG antibody. The domains and regions of the heavy (H) and light chains (L) are indicated. The variable (V) regions are at the amino-terminal ends and the constant (C) regions are at the carboxy-terminal ends. The interchain disulfide bonds and carbohydrate molecules are indicated.

Hybridoma Cells
Total RNA Extract
PCR V and C Regions for Heavy and Light Antibody Chains
PCR-SOE Gene Fragments Together
Digest with NcoI and XbaI
Ligate into Transfer Vector
Transform *E. coli*
Mini-prep Vector DNA
Sequence Cloned Gene
Transfect Insect Cells Using Linear Wild-type DNA and Vector DNA
(check cell pellet for protein synthesis by Western blot)
Harvest Supernatant and Plaque
Pick Plaques into 1 ml of Medium
Infect 2 x 10^5 Cells with 200 μl of Plaque Suspension
Harvest Cell Pellet for Western Blot Analysis
Prepare Purified Virus Stocks Producing H and L Antibody Chains to Passage 3
Coinfect insect Cells with H and L Chain Producing Viruses for Antibody Production
Harvest Supernatant and Purify Antibody
Demonstrate Antibody Production by Western Blot Analysis and/or ELISA
(Under Reducing and Non-Reducing Conditions)

Figure 2. Flow Chart For The Production Of Antibodies In Insect Cells

replaqued to ensure viral purity before generating virus stocks. In order to produce complete antibodies, we infect Sf9 cells with a MOI of 5-10 with both H chain and L chain viruses. We express only one antibody chain per virus, as opposed to having both H and L chain genes on the same virus. The reason for this is to maintain maximal flexibiliy to answer questions regarding the role of specific L or H chains involved in a particular biological interaction. Separate genes on each virus allows us to mix and match H and L chains at will.

The genes coding for the mouse and human antibodies that we are expressing in the baculovirus system have their natural leader sequences. The mouse signal peptide sequence for the 91A3 κ chain is 5' MVSSAQFLGLLLLCFQGTRC 3'; the 93G7 mouse gamma chain signal sequence is 5' MGWSFIFLFLLSVTAGVHS 3'; the human κ signal sequence used for all of our κ and λ light chains is 5' MEAPAHVLFLLLLWLPDTTG 3'; the human H chain signal sequence for the cold agglutinin FS-7 mu H chain is 5' MEHLWFFLLLVAAPRWVLS 3'. We have evidence indicating that Sf9 cells secrete more antibodies when the proteins are associated with the mouse signal sequences than with the human signal sequences. Antibody recovery improves even when the mouse signal sequence is used only on the L chain gene. We tested the effect of the acidic glycoprotein gp67 signal sequence 5' MLLVNQSHQGFNLGHTSLMVSAIVLYVLLAAAAHSAFA 3' (*32*) on immunoglobulin secretion. We do not find any increase in the amount of antibody recovered when the insect signal sequence is on the heavy chain alone, however, when it is on the light chain alone, more antibody is recovered from the supernatant than when the human signal peptide is used.. These findings indicate that the signal sequence can significantly influence the amount of antibodies secreted into the tissue culture medium.

Antibody Purification. Since human antibodies produced in the baculovirus system are entire molecules containing full-length V and C regions, these antibodies are subject to the purification methods applied to natural human antibodies. The IgG antibodies can be purified on protein A and protein G columns, while the IgM and IgA antibodies are generally concentrated from supernatants using Centriprep 100 or stirred ultrafiltration cells under positive pressure (Amicon, Beverly, MA). Ammonium sulphate and polyethylene glycol precipitation from serum containing media can also be used to concentrate antibodies with success, whereas it is very difficult to use the precipitation approach with serum-free medium due to inadequate precipitate formation. Antibodies with known specificities can be purified on affinity columns. While adequate amounts of material are harvested for analysis in a research laboratory, the concentration method is associated with substantial losses of antibody ranging from 20-80%. We have not added tags, such as hexahistidine or the strep tag (*48*), to our recombinant antibodies. We find that recombinant hexahistadine-tagged proteins produced in Sf9 cells are usually not sufficiently pure for biological analysis following single passage through a nickel column. When a second purification scheme is used in tandem with the nickel column, high purity can be achieved. Chimeric IgA antibodies having anti-arsonate activity in the V region and a human IgA C region have been affinity purified on an arsonate-Speharose column (*43*).

Antibody Characterization. Recombinant antibodies isolated from culture supernatants and infected cell pellets are analysed by Western blot on PVDF membrane (Immunobilon-P; Millipore Corp., Bedford, MA) and detected employing either the enhanced chemiluminescence protocol (Amersham Corp., Arlington Heights, IL) using horse-radish peroxidase conjugated anti-human IgM or IgG antibodies or alkaline phosphatase conjugated antibodies followed by nitro blue tetrazolium and 5-bromo-4-chloro-3-indolyl phosphate (GibcoBRL, Grand Island, N.Y). Recombinant antibodies are run on SDS-PAGE gels under reducing conditions to detect individual H and L chains and non-reducing conditions for the detection of assembled antibodies. Western blots demonstrate that Sf9 cells singly infected with IgG and IgM H chain encoding virus secrete barely detectable amounts of H chain peptides after concentration through YM 10 membrane (Amicon Corp., Beverly, MA), regardless of their associated signal sequences. However, L chain monomers and dimers are detected in the supernatant of singly infected, L chain producing cells. Following co-infection of Sf9 cells with H and L chain encoding viruses, complete four-chain heterodimers, as well as free heavy and light chains, L chain dimers and half-antibodies composed of one H chain and one L chain, are detected in culture supernatants. This indicates that the light chain is required for the majority of heavy chain secretion. Intracellularly, H and L chains are produced in copious amounts during both single and dual infections. Under non-reducing conditions there is a smear of H and L chain material, which upon reduction, produces single H and L chain bands. This indicates that intracellularly the immunoglobulin chains are forming disulfide bridges with cellular proteins other than their immunoglobulin partner chain. The human J chain, (gene obtained from Dr. E. Max, Bethesda, MD) appears to be glycosylated in Sf9 cells, as its migration is altered when produced in the presence of tunicamycin. Cells singly infected with J chain virus do not secrete J chain. However, J chain appears in the medium when cells are coproducing antibody H and L chains (*43*).

All of the IgG antibodies we have made in the baculovirus expression system are secreted as four-chain H_2L_2 monomers. The IgM antibodies are also predominanlty four-chain monomers, even in the presence of J chain. We found that some IgM antibodies will form multimers in the size range of dimers to pentamers, both in the presence and absence of J chain, although these multimers have not yet been characterized. Insects cells differ, therefore, from some non-lymphoid transfectants such as C6 glioma cells which secrete hexameric IgM (*49*). Antibodies of the IgA isotype are produced as monomers, but can form some dimers when coexpressed with J chain. In the presence of J chain, extracellular monomeric IgA is also associated with J chain (*43*).

The authenticity of recombinant antibodies is also analysed by testing the interaction with antibody specific monoclonal antibodies. These results confirm that, in general,recombinant antibodies produced in insect cells are indistinguishable (except for glycosylation) from antibodies isolated from human serum. Recombinant IgA antibodies reacted with a series of anti-IgA antibodies (*43*), and commercial preparations of anti-IgG, IgM and IgA all react with their respective antibodies as expected.

Glycosylation of Antibodies in Insect Cells. Antibodies produced in insect cells require glycosylation for secretion. In the presence of tunicamycin, cells coinfected with mu and L chain viruses produce their respective chains intracellularly, however, there is absolutely no antibody detected in the cell supernatants. IgA1 is a heavily glycosylated imunoglobulin containing seven carbohydrate chains, five of which are O-linked disaccharides coupled to serine residues in the hinge region between Fab and Fcα and two are complex oligosaccharides which are N-glycosylated linked to asparagines at residue 271 in C_H2 and residue 483 in the tail region (*50,51*). Results of a lectin binding study are consistant with the presence of high-mannose, N-linked glycosylation and O-glycosylation of IgA in Sf9 cells (*43*). Removal of the N-linked site at postion 271 had no effect on the secretion of IgA. Blocking N-linked glycosylation at both sites in the presence of tunicamycin, however, completely abrogated secretion. Lectin binding indicates that the recombinant IgM antibodies are associated with high-mannose N-linked glycosylation. In this respct, insect cells behave similarly with plasmactoma cells in that tunicamycin severely inhibits the secretion of the more heavily glycosylated immunoglobulins, IgA, IgM and IgE, but only has a modest effect on the secretion of IgG antibodies in these cells (*52,53*).

Production Levels. The amount of antibody produced is influenced by almost every component of the system. These include the transfer vector used, promoter, leader sequence, choice of insect cell and the medium composition. Quantitation of antibodies is done by ELISA using an antibody of known concentration as a standard. Antibodies are serially two-fold diluted and the concentration determined from the standard curve. We compared the amount of antibody produced by Sf9, Sf21, TN-368 and BTI-TN-5 cells. We found that there was no significant difference in the amount of human antibody produced among the three cells lines Sf9, Sf21 and TN-368. BTI-TN-5B1-4 cells, however, secreted 2-10 times more antibody than the other three types of cells. The range of antibody levels recovered from 80 ml of infected Sf9 serum-free cell supernatant by concentration to 1 ml in stirred ultrafiltration cells ranges from approximately 5 μg/ml to 17 μg/ml. Antibody levels in neat serum-free supernatant ranges from 0.3-1.5 μg/ml. Approximately 2-3 fold more antibody is produced when coinfected Sf9 cells are grown in serum containing medium.

Immunoglobulin heavy chain binding protein (BiP) (*54*) and protein disulfide isomerase (PDI) (*55*) are two proteins in the endoplasmic reticulum (ER) associated with the correct folding of proteins and their assembly into multimeric complexes. There is evidence that both proteins are involved in the correct intracellular assembly of immunoglobulins. The genes coding for murine BiP and PDI have been cloned into an insect transfer vector (*56*) and the effect of increased levels of these chaperone proteins on immunoglobulin assembly in insect cells is being investigated.

Applications of *in vitro* Produced Antibodies

Idiotypic Analysis. Antibodies, by virtue of being glycoprotein in nature, are themselves antigens. Individual antigenic determinants in the V region are referred to as idiotopes (Id). The sum of the idiotopes is called the idiotype. Analysis of idiotypes provides valuable information regarding immunoglobulin structure. We conducted a study to localize the 9G4Id, a marker for antibodies encoded by the V_H4-21 gene segment. Using the approach of exchanging sequences between two antibodies, one known to express the Id and the other known to be Id negative, we determined that the 9G4 Id is located in the FR1 of V_H4-21 encoded antibodies (*42*). Similarly, we also determined that the LC1 Id is located in the FR1 of V_H4 family gene segments other than V_H4-21 (*57*), and V_H6 encoded antibodies (unpublished data). These two studies indicate the ease with which this system produces mutant antibody chains and combinatorial antibodies with any desired pairing of H and L chain. In addition, one of the L chains used is from the mouse anti-arsonate antibody 91A3 and we have shown that it pairs well with human H chains.

Study of Antigenic Specificity and Superantigen Interactions. The structure of arsonate binding antibodies was studied by mutating an arsonate binding antibody (*58*). A series of recombinant antibodies was produced in the baculovirus system and tested for arsonate binding. It was found that L chains mutated at the V-J junction or H chains mutanted at the V-D junction were unable to bind arsonate. Mutations in the first and second CDRs of the H chain were shown to effect arsonate binding.

Staphylococcal protein A (SPA) is a member of a class of proteins called superantigens due to their ability to induce T-cell proliferation in the order of 5-40% compared with approximately 0.001% for conventional antigens. SPA is also able to bind determinants on the V regions of antibody H chains encoded by V_H3 gene segments (*59*). Antibodies produced in the baculovirus system were used to investigate the binding of SPA to the V region of V_H3 encoded antibodies (*60*). The parental IgM, a V_H3 encoded rheumatoid factor RF-SJ3, does not bind SPA. Based on a comparison of amino acid sequences between SPA binding and SPA non-binding antibodies, a single point mutation in the CDR2 at position 57 of RF-SJ3 was made and the mutant antibody efficiently bound SPA. This result localizes a critical site involved in SPA binding to the CDR2 of human immunoglobulins encoded by V_H3 family gene segments.

The Role of Glycosylation in Antibody Structure/Function. Antibodies are glycosylated in insect cells. Those of the IgA1 isotype have both O-linked Galβ(1-3)GalNAc disaccharides and N-linked high mannose oligosacccharides (*43*). Site-directed mutagenesis can be used to remove glycosylation sites to generate both glycosylated and non-glycosylated forms of the same antibody. It was determined by PCR-mutagenesis that N-glycosylation of the second domain of Cα is required for interaction with the monocyte IgA Fc receptor (*43*). This result demonstrates that N-linked glycosylation effects the conformation of effector binding sites. The effects of glycosylation on binding affinity with antigen can be studied using these antibodies. V_H4-21 encoded antibodies have a glycosylation site in the CDR2 of the V region, in addition to glycosylation sites in the C region. Removal of this V region site from the cold-agglutinin FS-7 does not prevent binding to the I antigen in an ELISA (unpublished data). Fine analysis of the effects of the removal of the glycosylation site will be determined using a plasmon resonance sensor.

The role of glycosylation in immunoglobulins is very much an area of active research (*61*). It is unlikely that insect cells, or any other cell type, will glycosylate antibodies in a manner identical to that of human B cells. Oligosaccaride processing is dependent on the presence of specific glycoslytransferases which are cell-type dependent. The effects of insect cell glycosylation on antibody structure, function and antigenicity will have to be determined for each antibody.

Constant Region Structure/Function Analysis. The flexibility of the system is demonstrated by the ability to join at the DNA level V and C regions which are not found in nature. We have produced a rheumatoid factor RF-TS1, which is naturally isolated as the IgM isotype, as IgG for purification purposes. Chimeric mouse-human IgA1 antibodies were produced in Sf9 cells as a means of mapping the sites of interaction of IgA receptors with the IgA1 constant region (*43*). The signal sequences and the V regions of the the H and L chains were the mouse anti-arsonate 93G7 and 91A3 sequences respectively attached by PCR-SOEing to the human alpha constant region. The recombinant antibodies bound the arsonate hapten, were recognized by monoclonal antibodies, fixed complement via the alternative pathway and specifically bound the monocyte IgA Fc receptor.

Production of Other Immune System Molecules. The human class II major histocompatibility complex (MHC) protein HLA-DR1 was produced in a membrane-bound and soluble, secreted form using α and β subunit genes (*62*). The subunits of

DR1 assembled as αβ heterdimers, however, the peptide-binding site was empty. This system potentially allows one to study the mechanism of peptide binding. Using another approach, two murine class II molecules were produced in soluble form in insect cells with a peptide attached by a linker to the N-terminus of their β-sheets. The peptide lies in the peptide-binding groove of the of the secreted MHC molecule and the complex is recognized by T cells having receptors specific for that particular combination (63). In another study, it was determined that the superantigen *Staphylococcus aureus* enterotoxin B (SEB) can form complexes with T cell receptor in both the absence and presence of class II MHC by using the HLA-DR1 produced in insect cells, (64). The cloning and expression of HLA-DR4Dw4 molecules produced in Sf9 cells has been reported (65) in amounts sufficient for biochemical and functional characterization. In addition, several cytokines and receptors have been expressed in the baculovirus system. Recombinant IL-2 was secreted into the culture medium and was able to stimulate growth of an IL-2-dependent cell line (66). Biologically active Il-5 (67) and Il-6 (68) have both been produced in insect cells. The yield of Il-5 from Sf9 cells was significantly greater than the levels produced in yeast cells. Lymphotoxin (LTα) and a soluble dimeric form of its receptor produced in Sf9 and BTI-TN-5B1-4 cells were both shown to be biologically active (69). The levels of receptor production in insect cells was approximately 2-3 mg/l over 5 days in BTI-TN-5B1-4 cells compared with approximately 0.05-0.1mg/l over 7 days in transiently transfected COS-7 cells. Human β interferon (70), human granulocyte-macrophage colony-stimulating factor (71), human erythropoietin (72) and the human 55 kD tumor necrosis factor receptor (73) have all been expressed in insect cells.

Future Directions

The combination of hybridoma technology together with the tools of molecular biology, such as the polymerase chain reaction (PCR) and DNA sequencing, allows for the production of tailored antibodies with respect to antigen specificity, H and L chain pairing, isotype and immunogenicity. It is now possible to produce human antibodies having specific idiotopic characteristics, and isotypic effector functions combined with defined antigen specificity in the baculovirus expression system. Large-scale production of antibodies is often the goal of research laboratories as well as for commercial purposes. The growth of insect cells as suspension cultures in stirred flasks or airlift fermentors in continuing to be investigated with the aim of economically producing large amounts of antibodies using insect cells (reviewed in 74, 75).

There is a role for complete antibodies in cancer detection using radiolabeled monoclonal antibodies (reviewed in 76). The technology is in a developmental stage dealing with issues such as affinity, immunogenicity and serum clearance. There are several clinical situations where complete antibodies are preferred to antibody fragments. Fab fragments are cleared faster from the serum, however, they have a tendency to be retained in the kidney for long periods. This limits the use of antibody fragments in detecting kidney disease. Antibody fragments are also more difficult to purify to clinical standards. Human antibodies produced in the baculovirus system could be used to address these concerns.

Acknowledgments

We acknowledge the National Institutes of Health (AI-12127 and CA44016) for financial support. J.D.Capra holds the Edwn L. Cox Distinguished Chair in Immunology and Genetics at the University of Texas Southwestern Medical Center at Dallas. We thank Brian E. Finley for assistance with the figure.

Literature Cited

1. Carayannopoulos, L.; Capra, J.D. In *Immunoglobulins, Structure and Function*; Paul, W.E., Ed.;Fundamental Immunology; Third Edition; Raven Press, Ltd.: New York, 1993, pp. 283-314.
2. Jerne, N.K. *Ann. Immunol.* (Paris) **1974**, 125C, 1127.
3. Wright, A.; Shin, S-U.; Morrison, S.L. *Crit. Rev. Immunol.* **1992**, 12,125.
4· Kabat, E.A.; Wu, T.T.; Perry, H.M.; Gottesman, K.S.; Fueller, C. *Protein sequences of immunological interest.* 5th ed.; U.S. Department of Health and Human Services, NIH publication no. 91-3242; 1991.
5· Hasemann, C.A.; Capra, *J. Proc. Natl. Acad. Sci. USA.* **1990**, 87, 3942.
6· zu Putlitz, J.; Kubasek, W.L.; Duchene, M.; Marget, M.; von Specht, B.-U.; Domdey, H. *BIO/TECHNOLOGY*. **1990**, 8, 651.
7· Laroche, Y.; Demaeyer, M.; Stassen, J-M.; Gansemans, Y.; Demarsin, E.; Matthyssens, G.; Collen, D.; Holvoet, P. *J. Biol Chem.* **1991**, 266, 16343.
8· Holvoet, P.; Laroche, Y.; Lijnen, H.R.; Cauwenberge, R.V.; Demarsin, E.; Broouwerst, E.; Matthyssens, G.; Collen, D. *J. Biol. Chem.* **1991**, 266, 19717.
9· Wu, T. T.; Kabat, E.A. *Mol. Immunol.* **1992**. 29, 1141.
10· Smith, G.E.; Fraser, M.J.; Summers, M.D. *J. Virol.* **1983,** 46, 584.
11· Summers, M.D.; Smith, G.E. Texas Agricultural Experimental Station Bulletin no. 1555, 1987.
12· O'Reilly, D.R.; Miller, L.K.; Luckow, V.A. *Baculovirus Expression Vectors, A Laboratory Manual*; W.H. Freeman and Company; New York, NY, 1992.
13· Fraser, M.J. *Current Topics in Micro. Immunol.* **1992**, 158, 131.
14· Kitts, P.A.; Ayres, M.D.; Possee, R.D. *Nuc. Acids Res.* **1990**, 18, 5667.
15. Ernst, W.J., Grabherr, R.M, Katinger, H.W.D., Nucleic Acids Res, 1994, 22, 2855.
16· Potter, K,N.; Faulkner, P.; MacKinnon, E. *J. Virol.* **1976**, 18,1040.
17· Miller, D.W.; Miller, L.K. *Nature* (London) **1982**, 299, 562.
18· Fraser, M.J.; Smith, G.E.; Summers, M.D. *J. Virol.* **1983,** 47, 287.
19· Beams, B.; Summers, M.D. *Virol.* **1990**, 174, 354.
20· Wickham, T.J.; Davis, T.; Granados, R.R.; Hammer, D.A.; Shuller, M.L.; Wood, H.A. *Biotechnol. Lett.* **1991**, 13, 483.
21 Kool, LM.; Voncken, J.W.; van Lier, F.L.J.; Tramper, J.; Vlak, J.M. *Virol.* **1991**, 183, 739.
22· Jarvis, D.L.; Garcia, Jr., A. *Bio/Techniques,* **1994,** 16, 508.
23· Lynn, D.E. *Bio/Techniques*, **1992,** 13, 282.
24 Shanafelt, A.B. *Bio/Techniques*, **1991**, 11, 330.
25· Vaughn, J.L.; Goodwin, R.H.; Thompkins, G.J.; McCawley, P. *In Vitro* **1977**,13,213.
26· Hink, W.F. *Nature,* **1970**, 226, 466.
27· Hink, W.F.; Strauss, E. In *Invertebrate Tissue Culture. Application in Medicine, Biology, and Agriculture*; Kurstak, E., Maramorosch K. Eds.; Academic Press Inc., New York, New York, 1976, pp 297-300.
28· Maiorella, B.; Inlow, D.; Shauger, A.; Harano, D. *BIO/TECHNOLOGY* **1988,** 6, 1406.
29· Rice, J.W.; Rankl, N.B.; Gurganus, T.M.; Marr, C.M.; Barna, J.B.; Walters, M.M.; Burns, D.*J. Bio/Techniques*, **1993**, 15, 1052.
30· Davidson, D.J.; Fraser, M..J.; Castellino, F.*J. Biochem.* **1990**, 29, 5584.
31 Thomsen, D.R.; Post, L.E.; Elhammer, A.P. *J. Cell. Biochem.* **1990,** 43, 67.

32· Iatrou, K.; Meidinger, R.G.; Goldsmith, M.R. *Proc. Natl. Acad. Sci. U.S.A.* **1989**, 86, 9129.

33· Gruenwald, S.; Heitz, J. Baculovirus expression vector system: procedures and methods manual, second edition; PharMingen, San Diego, CA. 1993.

34 Belyaev, A.S.; Roy, P. *Nuc. Acids Res.* **1993**, 21, 1219.

35· Peng, S.; Sommerfelt, M.; Logan, J.; Huang, Z.; Jiling, T.; Kirk, K.; Hunter, E.; Sorscher, E., *Prot. Express. Purif.* **1993**, 4, 95.

36· Ono, M.; Tucker, P.W.; Capra, J.D. *Nuc. Acids Res.* **1994**, 22, 3918.

37· Burand, J.P.; Summers, M.D.; Smith, G.E. *Virol.* **1980**, 101, 286.

38· Carstens, E.B.; Tija, S.T.; Doerfler, W. *Virol.* **1980**, 101, 311.

39. Potter, K.N.; Miller, L.K. *J. Invert. Pathol.* **1980**, 36, 431.

40 Mann, S.G.; King, L.A. J. Gen. *Virol.* **1989**, 70, 3501.

41 Goswami, B.B.; Glazer, R.I. *BioITechniques* **1991**, 10, 626.

42· Kitts, P. A.; Ayres, M.D.; Possee, R.D. *Nuc. Acids Res.* **1990**, 18, 5667.

43· Potter, K.N.; Li, Y-C.; Pascual, V.; Williams, R.C.; Byres, L.C.; Spellerberg, M.; Stevenson, F.K.; Capra, J.D. *J. Exp. Med.* **1993,** 178, 1419.

44· Carayannopoulos, L.; Max., E.E.; Capra, J.D. *Proc. Natl. Acad. Sci. USA* **1994,** 91, 8348.

45· Hasemann, C.A.; Capra, J.D. *Proc. Natl. Acad. Sci.* USA. **1990,** 87, 3942.

46· Horton, R.M;, Hunt, H.D.; Ho, S.N.; Pullen, J.K.; Pease, L.R. *Gene*, **1989, 77**, 61,

47 Ho, S.N.; Hunt, H.D.; Horton, R.M.; Pullen, J.K.; Pease, L.R. *Gene*, **1989, 77**, 51.

48· Schmidt, T.G.M.; Skerra, A. *Protein Engineering*, **1993**, 6, 109.

49 Cattaneo, A.; Neuberger, M.S. *EMBO J.* **1987**, 6, 2753.

50 Baenziger, J.; Kornfeld, S. *J. Biol. Chem.* **1974**, 249, 7260.

51· Torano, A.; Tzuzukida, Y.; Liu, Y.S.; Putnam, F.W. *Proc. Natl. Acad. Sci.* USA. **1977,** 74, 2301.

52· Hickman, S.; Kornfield, S. *J. Immunol.* **1978**, 121, 990.

53. Hickman, S.; Wong-Yip, Y.P. *J. Immunol.* **1979**, 123, 389.

54· Rothman, J.E. *Cell* **1989**, 59, 591.

55. Freedman, R.B.; Bulleid, N.J.; Hawkins, H.C.; Paver, J.L. B*iochem. Soc. Symp.* **1989,** 55, 145.

56· Hsu, T-A.; Eiden, J.J.; Betenbaugh, M.*J. Ann. New York Acad. Sci.* **1994,** 721, 208.

57· Potter, K.N.; Li, Y-C.; Capra, J.D. *Scand. J. Immunol.* **1994**, 40, 43.

58· Hasemann, C.A.; Capra, J.D. *J. Biol. Chem.* **1991**, 266, 7626.

59· Hillson, J.L.; Karr, N.S.; Opplinger, I.R.; Mannik, M.; Sasso, E.H. *J. Exp. Med.* **1993**, 178, 331.

60· Randen, I.; Potter, K.N.; Li, Y-C.; Thompson, K.M.; Pascual., V.; Forre, O.; Natvig, J.B.; Capra, J.D. *Eur. J. Immunol.* **1993**, 23,2682.

61· Wright A.; Morrison, S.L. *J. Exp. Med.* **1994**, 180, 1087.

62· Stern, L.J.; Wiley, D.C. *Cell* **1992**, 68, 465.

63· Kozono, H.; White, J.; Clements, J.; Marrack, P,; Kappler, *J. Nature* **1994,** 369, 151.

64· Seth, A.; Stern, L.J.; Ottenhoff, T.H.M.; Engel, I.; Owen, M.J.; Lamb, J.R.; Klausner, R.D.; Wiley, D.C. *Nature* **1994,** 369, 324.

65· Scheirle, A.; Takacs, B.; Kremer, L.; Marin, F.; Sinigaglia, F. *J. Immunol.* **1992**, 149, 1994.

66. Smith, G.E.; Ju, G.; Ericson, B.L.; Moschera, J,; Lahm, H-W.; Chizzonite, R.; Summers, M.D. *Proc. Natl. Acad. Sci. U.S.A.* **1985**, 82, 8404.
67· Ingley, E.; Cutler, R.L.; Fung, M-C.; Sanderson, C.J.; Young, I.G. Eur. *J. Biochem.* **1991**,196, 623.
68· Matsuura, Y.; Tatsumi, M.; Enami, K.; Morikawa, S.; Yamazaki, S.; Kohase, M. Lymphokine, *Cytokine Res.* **1991**, 10, 201.
69· Crowe, P.D.; VanArsdsale, T.L.; Walter, B.N.; Dahms, K.M.; Ware, C.F. *J. Immunol. Methods* **1994**, 168, 79.
70· Smith, G.E.; Summers, M.D.; Fraser, M.*J. Mol. Cell. Biol.* **1983**, 3, 2156.
71· Chiou, C-J.; Wu, M-C. *FEBS Letters* **1990**, 259, 249.
72 Wojchowski, D.M.; Orkin, S.H.; Sytkowski, A.*J. Biochim. Biophys. Acta.* **1987**, 910, 224.
73· Loestscher, H.; Pan, Y-C. E.; Lahm, H-W.; Gentz, R.; Brockhaus, M., Tabuchi, H., Lesslauer, W. *Cell* **61**, 351.
74 Taticek, R.A.; Lee, C.W.T.; Shuler, M.L. *Current Biol.* **1994,** 5, 165.
75· King, G.A.; Daugulis, A.J.; Faulkner, P.; Goosen, M.F.A. *Biotechnol. Prog.* **1992**, 8, 567.
76 Divgi, C.R. *Cont. Oncol.* **1994,** 4,5 6.

RECEIVED May 17, 1995

Chapter 5

Antibody Expression in Plants

J. K-C. Ma

Department of Immunology, United Medical and Dental Schools of Guy's and St. Thomas's, Guy's Hospital, London SE1 9RT, England

The development of reliable techniques for transformation of plant cells has led to the possibility of expressing a range of proteins in plants. Of these, expression of antibodies is perhaps the most interesting, because there are so many potential applications. Plants could be used as bio-reactors for large scale antibody production, alternatively, antibodies could be used to modulate plant biochemistry, or influence disease resistance. Antibody processing in plants is very similar to that found in mammalian plasma cells. Signal sequences direct the light and heavy chains to the endoplasmic reticulum for assembly and glycosylation and the assembled antibody is secreted, via the Golgi apparatus. The potential for antibody engineering in plants is high. Modifications to the heavy chain, including the extension of the heavy chain by one domain, can be accomodated without affecting the fidelity of assembly. Antibody fragments can also be produced which may be targeted for secretion, or for intracellular expression by removal of the signal sequence. Secreted antibody accumulates in the extracellular apoplastic fluid, which is a vast, but stable environment, and so levels of recovery can be extremely high.

Genetic engineering of plants has been practised for centuries, in order to introduce desirable traits into crops and other plants. Over the last 10 years however, the science has advanced by leaps and bounds, following the development of simple and reliable molecular biology techniques for introducing genes into plant cells. Much of the research is focused on strategies for improving the performance of plants, but the potential for using plants as bio-reactors or for producing bioactive compounds for human use, is also of enormous interest.

There are a number of advantages to using plant systems for genetic engineering. Firstly, plants have an enormous regenerative capacity. Many plant cells can readily be propagated in culture as protoplasts (cells from which the cell walls have been removed), or as undifferentiated callus cells, which can be propagated in liquid suspension or on agar. From these cell cultures, which can be used for transformation procedures, it is then possible to regenerate whole plants. Secondly, as described later, most plant transformation techniques result in the stable integration of the foreign DNA into the plant genome. Thus, genetic recombination by sexual

0097–6156/95/0604–0056$12.00/0

crossing of transgenic plants is a simple method to introduce new genes into plants, or plants homozygous for a particular gene can readily be generated by self fertilisation. Furthermore, the transformed plant line can then be stored almost indefinitely as seeds, under ambient conditions. Finally, once a transgenic plant line has been established for a chosen recombinant protein, the potential yield is enormous, as production could be scaled up to agricultural proportions. The cost of such production would therefore be extremely modest.

In 1989, the production of a murine antibody in transgenic plants was first described (1). The finding that the antibody was fully functional, demonstrated that plant cells could assemble a mammalian multimeric protein correctly. In comparison with other expression systems for antibodies, it is important that plants can produce full length antibody, that is glycosylated, at high levels of expression. It has also been shown that plants can produce antibody fragments.

This chapter will cover various aspects of the expression of antibodies in plants. It will review methods used for transformation of plant cells, intracellular assembly, post-translational processing and secretion of the plant antibody, as well as the potential for antibody engineering in plants. Finally, future applications for plant antibodies will be discussed.

Plant Transformation.

In the most widely used technique for plant transformation, *Agrobacterium tumefaciens* acts as a delivery vehicle for the introduction of recombinant plasmid vectors into the plant cell (2). In nature, *A. tumefaciens* is a soil borne bacterium that causes the neoplastic, crown gall disease. The agrobacterium never actually enters the cell, but responds to chemical signals released by wounded cell walls to elicit the transfer of genetic material. A number of genes that are contained in a 20Kb section (T-DNA) of a circular extra-chromosomal tumour inducing (Ti) plasmid are inserted into the plant genome. The integration of the T-DNA is not sequence specific, and requires only a region of DNA from the border of the T-DNA/Ti plasmid junction (3). Thus, any plasmid DNA that includes this border region, can be introduced into the plant cell nucleus and stably integrated into the plant nuclear genome. Disarming the plasmid by removal of the virulence genes involved in neoplastic transformation does not prevent integration, and they can be replaced by heterologous DNA along with promoters that are active in plant cells that will direct the expression of the transgene. Integration of this modified T-DNA will result in transformed plant cells devoid of neoplastic characteristics, which can be regenerated into mature fertile plants.

The vectors that have been developed, contain selectable plant markers, promoters upstream from polylinkers and *E. coli* and agrobacterium origins of replication (4). These vectors are generally quite large, and the initial strategy to produce antibody in plants was to express each immunoglobulin chain separately in different plants and to introduce the two genes together in the progeny plant by cross pollination of the individual heavy and light chain expressing plants (Figure 1).

Not all plants are amenable to transformation by agrobacterium, notably the monocotyledonous plants. Other techniques are available for introducing DNA into plant cells, including the use of viral vectors (5) or by shotgun microprojectile bombardment (6). By one technique or another, many plant species, including most of the important crop plants can now be transformed.

Practical aspects. The generation of transgenic plants is not a rapid procedure. From transformation of the plant cell to regeneration of a mature plant, can take 2-3 months in the case of tobacco. The time from flower fertilisation to seed collection is approximately 4 weeks, and from sowing of seed to a plantlet ready for assaying, a further 2-3 weeks. Thus the entire process of generating a homozygous plant line transgenic for a single recombinant protein can take 4-5 months. For full length antibodies, the approach we have taken, as described above, involves 2 generations of plants to generate an antibody producing plant (1,7). Using this strategy, the yield of recombinant antibody is consistently high, between 1-5% of total plant protein. Others have incorporated both the light and heavy chain genes into the

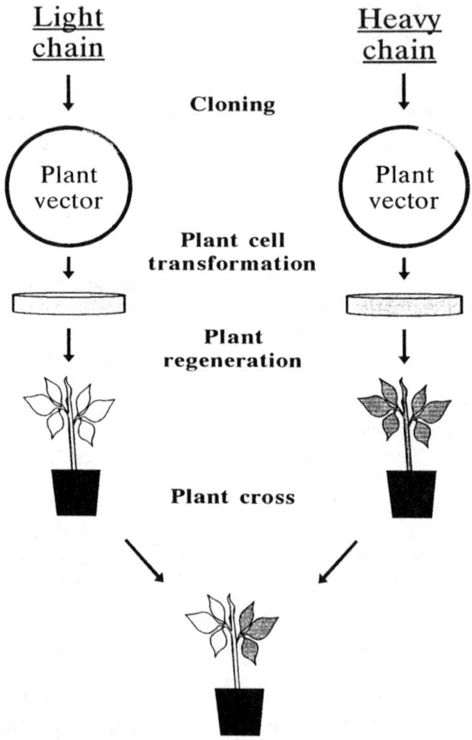

Figure 1: Strategy for production of antibody in tobacco plants. Plants are regenerated following transformation with agrobacterium containing antibody heavy or light chain cDNA. These are sexually crossed to produce progeny which secrete both chains and assemble functional antibody.

same construct for transformation (8,9). This reduces the time required to produce an antibody secreting plant but the transformation may be less than optimal. Using this technique, the reported yield of antibody was low, amounting to only 0.055% of total soluble protein and the efficiency of assembly also appeared to be reduced (9). In the same study, a comparison between two plant species was made, which demonstrated that accumulation of assembled Fab was significantly greater in arabidopsis, as compared with tobacco, using the same construct for transformation. The result suggests that different plant species may be more or less suitable for antibody production and that tobacco, although a widely used plant model, may not be the ideal choice.

Processing of Antibodies in Plants.

Assembly of Full Length Antibodies. Several groups have reported the production of full length antibodies in plants (1,7-9). The antibodies are functional in terms of antigen recognition and binding, which is a critical and sensitive test that correct assembly of heavy and light chains has occurred. In mammalian plasma cells, the mechanism of assembly is only partially understood. The immunoglobulin light and heavy chains are synthesised as precursor proteins, and the signal sequences direct translocation into the lumen of the endoplasmic reticulum. Within the ER, the signal peptides are cleaved and at least two stress proteins, BiP/GRP78 and GRP94, function as chaperones which bind to unassembled heavy and light chains and direct folding and assembly within the endoplasmic reticulum (10,11). In plants, the same pathway is utilised and assembled antibody has been detected within the ER by electron microscopic immunogold labelling (8). Hiatt et al. demonstrated the need for signal peptides to direct the nascent light and heavy chains to the ER (1). Assembly of antibody occurred only if both chains were synthesised with a signal sequence, and omission of this sequence from either chain resulted in failure of assembly. The origin of the signal sequence did not appear to be critical, as plant and mouse sequences, as well as a yeast pro-sequence have been used successfully (1,8,12). Within the ER, evidence for proteolytic processing of the signal sequence had occurred correctly, was obtained by N-terminal sequencing of the light chain (12).

Chaperones homologous to BiP and GRP94 have been described in maize (13) and barley (14). b-70 is a member of the hsp70 family that is expressed during maize seed maturation and it is endosperm specific. Purified b-70 fractions contained 2 species of 75Kd polypeptides. Like BiP, both species were located in the ER and bound to ATP. One of the polypeptides was also modified post-translationally in a similar manner to BiP, namely by phosphorylation and ADP-ribosylation. The expression of b-70 was associated with the accumulation of an abnormal protein which suggests a functional similarity with BiP. Expression of the GRP94 homologue identified in barley was associated with stress induced by infection by the powdery mildew fungus or heat shock. It was found to possess a characteristic N-terminal domain, including a signal sequence, as well as the C-terminal KDEL ER retention sequence.

The mechanism of antibody assembly in plants has yet to be determined, but the presence in plants of chaperonins homologous to those found in mammalian cells, suggests that similar mechanisms could operate. Indeed, it would be surprising if the mechanisms controlling protein folding and assembly in plants and mammals were not broadly similar. Thus, it is consistent with the mammalian model for light and heavy chain assembly, that assembly of a complete tetrameric antibody entails an increase in protein accumulation, indicating that the native conformation of immunoglobulin molecules is an important determinant of stability (1).

Glycosylation. Protein modification by N-linked glycosylation is found in all higher eukaryotes. Many plant proteins contain N-linked as well as O-linked glycans and indeed, N-linked glycosylation in plants is similar to that in mammals (15-17). Thus, synthesis of the core high mannose oligosaccharide occurs in the ER, a process that is catalysed by several highly specific glycosyltransferases (18). The core high mannose type glycans have identical structures in plants, mammals and other organisms. They can be modified subsequently in the Golgi apparatus by the actions of a number of glycosidases and glycosyltransferases, to complex glycans (15,19). Complex plant glycans may be quite heterogeneous but tend to be smaller than mammalian complex glycans. They also differ in the terminal sugar residues, which can consist of xylose, fucose, N-acetylglucosamine, galactose, arabinose and rhamnose (15,16,20). A xylose residue linked β1-2 to the β-linked mannose residue of the glycan core is frequently found in plants and invertebrates, but not mammals (16). By comparison, in mammals, a predominant terminal residue is N-acetyl neuraminic acid (NANA) (18), a carbohydrate that has not been identified in plants.

A detailed analysis of the glycans associated with plant antibodies has not been performed, but a preliminary comparison between plant and murine derived antibodies has been made using lectin binding studies (12). Both types of antibody were bound with similar affinity by concanavalin A (specific for mannose and glucose). Prior treatment with endoglycosidase H did not appear to affect Con A binding, suggesting that the glycans from both antibodies had been processed in a similar fashion to complex type glycans. Lectins recognising N-acetylglucosamine oligomers, N-acetyl-lactosamine or galactose (β1,4) N-acetylglucosamine (β1,2) mannose did not bind to either antibody, but the lectins from *R.communis* agglutinin, which can bind terminal galactose and N-acetylgalactosamine and wheat germ agglutinin, which can bind terminal NANA recognised only the murine antibody. Thus the results suggest a distinct composition of the terminal residues in the plant glycan, and the differences were consistent with the absence of NANA in plants.

These differences in glycosylation patterns had no effect on antigen binding or specificity. However, in the case of antibodies produced in plants for human therapy, the presence of plant specific glycans might increase the immunogenicity of the recombinant antibody. If so, the complex glycans might be removed enzymatically, or the heavy chain sequence altered, to remove the Asn-X-Ser/Thr motif sequence, which is the site for N-linked glycosylation. An alternative, more elegant approach could also be adopted, by the use of mutant plants lacking enzymes involved in the glycosylation pathway (21).

Secretion and Targeting of Antibody. As in mammalian cells, following assembly and post-translational processing, antibodies are secreted by plant cells (12). This was demonstrated by pulse-chase labelling experiments, using protoplasts isolated from regenerated transgenic plants, to demonstrate the secretion of the antibody through the cell membrane. Furthermore, a similar experiment using callus cells, also demonstrated that antibody secretion occurs across the plant cell wall. This latter finding was surprising, as the pore size in the cell wall was previously thought to have an upper exclusion limit equivalent to a 20Kd protein (22). However, a small population of larger pores may also exist, which would allow passage of larger proteins and carbohydrates (22).

Accumulation of secreted antibody into the apoplasm could be advantageous. The apoplastic space is a large aqueous space external to the cells, in which there is minimal hydrolytic processing. Thus the antibodies are secreted into, and can accumulate within a stable environment. Firek et al., have shown that accumulation levels of an single chain Fv antibody fragment are considerably higher if the scFv is targetted for secretion, rather than expressed within the cytoplasm (23). An additional advantage is that extraction of antibodies from apoplastic fluid is simple and can be achieved under more mild conditions than those required for proteins located elsewhere.

Extracellular secretion is not the only option for full length antibody. We have also found that by the addition of membrane spanning and intracellular regions derived from the murine membrane Ig receptor, recombinant full length antibody can be retained within the cell membrane (Ma et al., MS in preparation). Targeting antibody expression at a cellular level, may be an important consideration, depending on whether the antibody is required for "intracellular immunisation", or disease resistance, or if its accumulation and storage is required for large scale production.

The apparent misdirection of assembled antibody into chloroplasts has been reported (8). In this study, light and heavy chains of an IgM antibody were co-expressed using a plant signal peptide derived from barley α-amylase. Assembly of the functional antibody was detected in the ER, but although the α-amylase signal sequence is derived from an extracellular protein of barley aleurone, cytolocalisation studies indicated the accumulation of the antibody in chloroplasts and ER, but not in the cell wall or intracellular spaces. The failure of secretion may have been related to processing of the barley leader sequence. It may also be possible that antibody assembly can occur in the chloroplast. Although the BiP and GRP94 chaperonins are only found in the ER, other chaperone molecules of the HSP70 family are found in chloroplasts (24), as well as HSP60 chaperonins, including the Rubisco binding protein (25). As all these chaperone molecules are thought to function in the folding and assembly of proteins, they may be able to perform a similar assembly function for immunoglobulin chains.

Most groups have used the constitutive 35S promoter from cauliflower mosaic virus (26,27). With this promoter, all plant cells would be expected to express the recombinant protein, although there is some evidence for differential levels of expression between different tissues. It should be possible in future however, to direct tissue specific expression, for example, to seed, fruit or leaves, depending on the particular application designed for the antibody.

Antibody Engineering In Plants.

It is possible to engineer all forms of antibody in plant cells, including antibody fragments and single chain molecules, as well as design modifications of the constant regions of the immunoglobulin. As with other expression systems, the recombinant protein can also be targetted for intracellular and extracellular expression.

Antibody Fragments. A wide range of functional recombinant antibody fragments have been described (28-32). Fab, Fv, single chain and single domain antibodies can all be readily produced in most heterologous systems and compared with full length antibody, the requirements for assembly are thought to be less demanding. As targeting to the ER is not essential, the antibody fragments can accumulate intracellularly if required, or if secretion into the extracellular space is preferred, this can be achieved by the inclusion of a signal sequence. In either case, no glycosylation occurs, because the normal antibody glycosylation site is absent.

Most of the antibody fragments described in *E.coli* have also been produced in plants. Benvenuto et al., have expressed a single domain antibody (dAb) in tobacco (33), and others have reported single chain Fv molecules (34,35). De Neve has also examined Fab production in tobacco and arabidopsis (9). Interestingly, in this study, comparative accumulation levels of assembled Fab or whole IgG of the same antibody were similar, despite the supposed differences in complexity of assembly.

Full Length Antibody. In retaining the constant region of the antibody molecule, many of the functional regions that are not involved in antigen recognition are preserved. These include the sites for glycosylation, complement binding, phagocyte binding, the hinge region, as well as the site for association with J chain and secretory component (in α and μ chains). Moreover, the avidity of binding and the ability to bind antigen bivalently (and therefore to cause aggregation) is maintained.

We have recently been investigating the production of full length monoclonal antibodies in transgenic tobacco plants, using either the original gamma heavy chain or redesigned hybrid heavy chains. The latter consist of the native variable domain and gamma constant domains, coupled to alpha chain constant domains, to result in a chimaeric IgA construct of the parent IgG antibody. This approach was used to determine whether redesigning the constant region of the antibody would affect the assembly or function of the hybrid antibody in plants.

The parent antibody (Guy's 13) is a murine IgG1, which recognises a cell surface adhesin (streptococcal antigen I/II) of *Streptococcus mutans*. The antibody has a potential use in immunotherapy against dental caries (36). As well as the original γ1 heavy chain, two types of chimaeric heavy chain were made by ligating fragments of DNA amplified by PCR from the Guy's 13 γ1 chain and an α chain from the MOPC 315 hybridoma (Figure 2). As compared with the native heavy chain, the G1/A heavy chain has Cα2 and Cα3 domains replacing the original Cγ2 and Cγ3 domain. In the G2/A heavy chain, the original Cγ1 and Cγ2 domains are present but the Cγ3 domain is replaced with Cα2 and Cα3 domains making this heavy chain longer than usual by one domain. The heavy chain DNAs were introduced separately into a plant vector downstream from the DNA encoding the native Guy's 13 leader sequence and under the control of the CaMV 35S promoter. Plant cells were transformed and transgenic plants regenerated and then crossed with a transgenic plant expressing native Guy's 13 light (kappa) chain. Thus, 3 different forms of Guy's 13 MAb were expressed in the progeny plants all containing the identical light (kappa) chain, but different heavy chains and these are referred to as Plant G13 (original γ1 heavy chain), Plant G1/A and Plant G2/A. In all experiments, the Guy's 13 hybridoma cell culture supernatant (referred to as Mouse G13) was used as a positive control. Negative control plants were those that had been transformed with the plant vector containing an insert that encoded an irrelevant mouse protein.

Western blot analysis of plant extracts was used to confirm the expression of antibody light, heavy, and chimaeric heavy as well as assembled antibodies in transgenic plants. Under reducing conditions, the presence of light (kappa) chain at approx. M_r 25K was detected in the mouse Guy's 13, as well as in the three transgenic plants, but not in the control plant (not shown). Guy's 13 heavy (gamma) chain was also detected in Plant G13 at approx. M_r 57K, but not in the control plant extract (not shown). The heavy chains of Plant G1/A and G2/A were detectable using an anti-alpha chain antiserum (not shown). Compared with the mouse Guy's 13 heavy chain, (approx. M_r 57K), the heavy chain of Plant G1/A had a slightly higher relative molecular mass (approx. M_r 60K) and the Plant G2/A heavy chain is much larger (approx. M_r 70K). This was consistent with the molecular weights predicted by sequence analysis. The anti-alpha chain antiserum did not cross react with the mouse Guy's 13, which only contains gamma chain domains.

Samples were also run under non-reducing conditions to confirm the assembly of heavy and light chains into an immunoglobulin molecule (Figure 3). Detection was with a labelled anti-kappa antiserum, and all three transgenic plants had assembled immunoglobulin at the correct M_r of above 150Kd for full length antibody. The Plant G13 antibody (lane b) has the same M_r as the mouse G13 (lane a), but the Plant G2/A (lane c) and Plant G1/A (lane d) antibodies have higher M_r as predicted. A number of smaller proteolytic fragments were also detected, which is consistent with previous findings (1), and that a number of proteases are released by plants during the antibody extraction procedure. That these are antibody fragments, is confirmed by the absence of any detectable bands in the control plant extract (lane e).

ELISA was used to detect the binding function of the assembled antibody to streptococcal antigen I/II (Figure 4). All the antibody solutions were adjusted to an initial concentration of 1.5μg/ml and used in serial 2-fold dilutions. Extracts from plants expressing either Guy's 13 heavy or light chain singly, were also included in these assays, to determine if the single immunoglobulin chains exhibited any antigen binding activity. Specific antibody was detectable in all three transgenic plant extracts, and the titration curves were similar to that of the murine hybridoma cell culture supernatant, used at the same concentration. The binding of the Plant G1/A antibody appeared to be slightly lower than the other antibodies, although the titration curve followed a similar pattern. No SA I/II binding activity was detected in the negative control plant nor did extracts from plants individually expressing light or heavy chains have binding activity to purified SA I/II. These findings demonstrate that the transgenic plants expressing both light and heavy chains have assembled the antibody molecule correctly to form a functional antigen binding site and that single light or heavy chains are not capable of binding the antigen.

The plant antibodies also recognised native antigen on the surface of streptococcal cells (*S. mutans* serotype c), which further confirms the integrity of the antigen binding site in the plant antibodies (not shown). There was no binding to *E. coli* cells by any of the plant extracts, at concentrations of 1.0 and 0.5 μg/ml.

It was important to confirm that the full length antibodies produced in plants were still able to bind antigen bivalently. This was demonstrated by incubation of *S.*

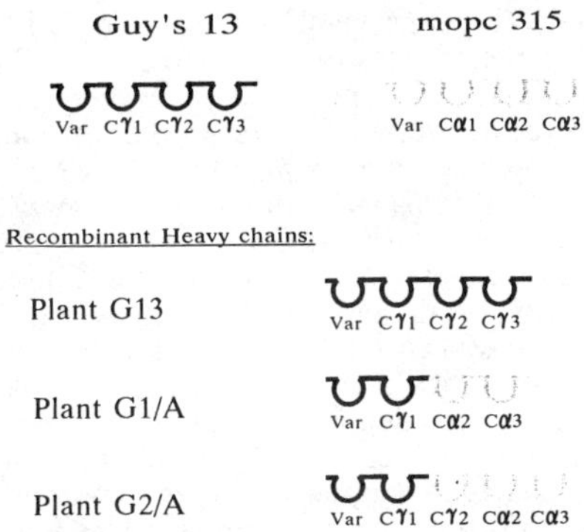

Figure 2: Recombinant heavy chains introduced into transgenic plants. Native Guy's 13 heavy chain (Plant G13) was used, as well as hybrid IgG/A heavy chains constructed by the replacement of 1 or 2 constant region domains from Guy's 13 with 2 domains from MOPC 315, a mouse IgA monoclonal antibody.

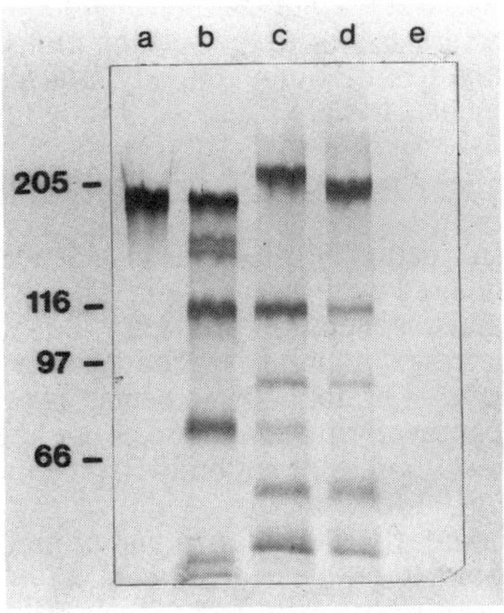

Figure 3: Western blot analysis of transgenic plant samples under non-reducing conditions. Detection was with a Goat anti-mouse kappa chain anti-serum, followed by alk. phosphatase labeled Rabbit anti-goat IgG antiserum. Samples are: a-mouse Guy's 13; b-Plant Guy's 13; c-Plant G2/A; d-plant G1A; e-control plant. (Reproduced from ref. 7. Copyright 1994 VCH Verlagsgesellschaft mbH.)

mutans cells with the antibody preparations in order to elicit bacterial aggregation. Plant extracts were sterilized by filtration through a 0.22μm pore size filter and diluted ten-fold with Todd Hewitt broth. Samples were inoculated with 0.05 vol. of an overnight *S. mutans* culture and incubated at 37°C overnight. The samples were gram stained and examined under oil-immersion microscopy. *S. mutans* grown in the presence of mouse Guy's 13 (Figure 5, panel A), plant Guy's 13 (panel B), plant G1/A (panel C) or plant G2/A (panel D) became aggregated and cell clumping was evident. However, the control plant extract had no effect on *S. mutans* growth (panel E).

The results have shown that by a modification of the cloning procedure, it was possible to replace γ constant region domains of the Guy's 13 heavy chain, with the corresponding constant region domains from an IgA α heavy chain and express the engineered heavy chain in plants. Furthermore, when co-expressed with light chain, the immunoglobulin chains were assembled into functional antibody. Thus, functional regions of an IgA antibody (Cα2 and Cα3), such as the J chain and secretory component binding sites, have been incorporated into the Guy's 13 MAb, and the results show that this did not impair antigen recognition or binding. It is possible that the plant technology could be extended to assemble a multimeric version of Guy's 13, in which the increased valency and resistance to proteolytic activity may be advantageous, especially if bacterial aggregation is an important effector mechanism.

The plant G2/A construct heavy chain contained an extra CH2 domain, compared with conventional antibodies and had a correspondingly higher molecular weight (Figure 3). The additional domain did not affect the fidelity of antibody assembly or antigen recognition and the antibody was functional. In the case of Plant G2/A, retaining the γCH2 domain preserves a protein A binding site, which may simplify purification of the antibody. However, for other antibodies, there might be significant advantages in incorporating other functional regions, such as the complement binding domain from a human IgG1 antibody, or domains containing regions that act as receptors for cell or tissue specific molecules. Although the ability to bind antigen is an essential feature of an antibody, it is also important to remember that most of the effector mechanisms are mediated through the constant regions.

Applications Of Monoclonal Antibodies Expressed In Plants.

As discussed earlier, transformation of plants to produce monoclonal antibodies can be a lengthy process, compared with other techniques. There will undoubtedly be improvements in vectors and regeneration protocols, but the technology is certainly better suited to well characterised antibodies, produced for a specific purpose, rather than studies in which some level of antibody screening is still necessary. There are 5 broad areas in which applications for plant antibodies are likely to be focused:

1) Mass production of antibodies or antibody fragments,
2) Passive immunisation,
3) Production of antibodies for which functional or modified Fc are required,
4) Improvement of plant and crop performance,
5) Immunomodulation for research into plant metabolism and biochemistry.

Mass Production Of Antibodies Or Antibody Fragments. The prospect of harvesting monoclonal antibodies on an agricultural scale offers the potential for extremely economic production of almost limitless amounts of antibody. It has been estimated, based on 1990 costs, that by expressing antibody in soybean at a level of 1% of total protein, 1kg of antibody could be produced for approximately US $100 (37). This is at least 1000-fold less than the cost of production by conventional hybridoma techniques. Many antibody applications do not require large amounts of material. However, as the capacity for production increases and the cost decreases, the number of possible uses will certainly extend into areas which at present may not even be considered, such as industrial scale purification processes.

Passive immunisation. At present, one area which requires large quantities of antibody is the *in vivo* application of therapeutic antibodies. Monoclonal antibodies have been used in human diseases with varying degrees of success, and a consistent

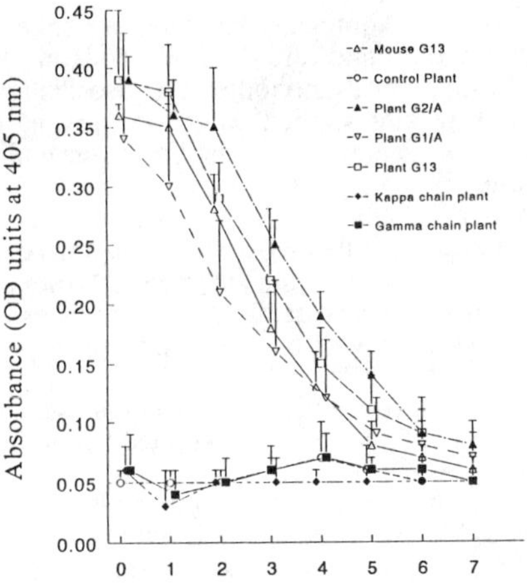

Figure 4: Binding of antibody samples to purified streptococcal antigen in ELISA. Triplicate samples were added in two-fold dilutions ad the results are shown as mean absorbance ± SD from three separate experiments. The starting concentration of each antibody solution was 1.5 µg/ml. (Reproduced from ref. 7. Copyright 1994 VCH Verlagsgesellschaft mbH.)

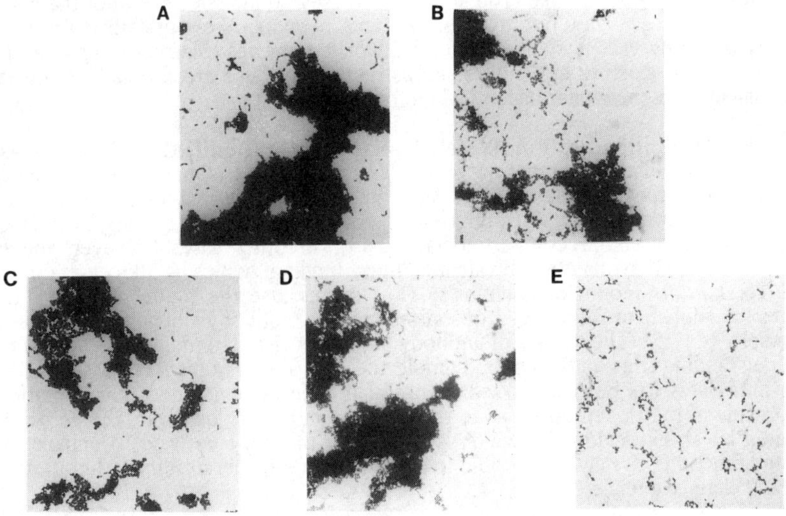

Figure 5: Aggregation of *S. mutans* following incubation with plant antibodies. S. mutans were grown in Todd Hewitt broth overnight at 37°C with 10% transgenic plant extract. Gram stained samples are shown in Panels A- Mouse Guy's 13; B- Plant Guy's 13; C-Plant G1/A; D-Plant G2/A; E-Control plant. (Reproduced from ref. 7. Copyright 1994 VCH Verlagsgesellschaft mbH.)

problem arises from the need to administer sufficient amounts of antibody at the site of disease, and to overcome the rapid rate of clearance from the body. As described earlier, topically applied monoclonal antibodies are also being considered for human use and the efficacy of topically applied MAb to prevent oral disease has been demonstrated in the case of dental caries, in both sub-human primates (38) and humans (36). Dental caries is mainly caused by a bacterium *Streptococcus mutans*, which colonises the teeth of children and later becomes established as part of the commensal oral flora. Topically applied MAb, that was raised against the cell surface adhesin of *S. mutans* prevented the establishment of the bacteria in non-human primates, and also reduced the levels of disease. In humans, the MAb was applied directly to the teeth and were shown to confer long term protection against *Strep. mutans* in adults. Although the optimal dose of MAb has yet to be determined, in order to overcome the significant dilution effect of saliva, in the human trial approximately 2.5mg of purified antibody was applied to the teeth on 6 occasions per individual.

The use of plants for orally delivered antibodies is another attractive possibility. It may be possible to produce these antibodies in edible plants or fruit, which would reduce or possibly eliminate the need for purification of the plant antibody. Indeed, antibodies against *Strep. mutans* have also been delivered topically in food and were shown to prevent colonisation and disease (39-41).

Plants also have a number of other advantages with respect to storage and distribution, which are pertinent to antibody based vaccines. Plant genetic material is readily stored in seeds, which are extremely stable and require little or no maintenance. Seeds have an almost unlimited shelf life in ambient conditions, unlike the more stringent requirements of bacteria or mammalian cells. Immortalisation of the plant line therefore is extremely simple and convenient and furthermore, mature plants can be self fertilised to produce identical offspring. A further advantage is that as the number of transformable crop species increases, to include several of the major food crops such as rice, potato, cassava and peppers, the possibility arises of delivering antibody vaccines cheaply to third world countries using indigenous crops. In this way, the existing agricultural infrastructure in these areas could be utilised rather than building new pharmaceutical factories.

Production Of Antibodies For Which Functional Or Modified Fc Regions Are Required. For many therapeutic antibodies, the emphasis has been placed on producing small functional antibody fragments, in order to minimise the antigenic challenge on immunization, to gain tissue penetration (31) or to target antibody fragments to intracellular compartments (42,43). In other cases however, the Fc region is required to maintain bivalent antigen binding, which is the case for the anti-*Strep. mutans* monoclonal antibody (36) or because the mode of action also involves complement binding, opsonisation or antibody dependent cellular cytotoxicity (44,45). The avidity of antibody binding is also determined by the number of antigen binding sites on the antibody molecule. The ability of plants to assemble full length antibodies has been discussed. It has also been demonstrated that alterations can be made in the Fc region without affecting assembly. For example, the Fc can be elongated by an extra domain, which is a strategy that may be used to enhance Fc mediated functions. It may be possible to accommodate further designed alterations in the heavy chain sequence.

Improvement Of Plant And Crop Performance. Antibodies or antibody fragments can be expressed in the cytoplasm of plant cells, in chloroplasts, in the cell membrane, or in the apoplastic space. Although antibody mediated protective mechanisms may be limited in plants, recombinant antibody molecules can be effective in insect or disease resistance. Tavladoraki et al. have recently reported the cytoplasmic expression of a single chain Fv in transgenic tobacco which protected against infection by artichoke mottled crinkle virus (35). Transgenic plants had significantly lower levels of virus accumulation, a reduced incidence of infection (approx. 50-60%) and a marked delay in the appearance of symptoms compared with control plants.

Voss et al have also reported the production of transgenic tobacco expressing a full length antibody specific for tobacco mosaic virus. These plants showed a reduction in their susceptibility to viral infection, and the reduction in symptoms was correlated with the amount of antibody produced in individual transgenic plants (Abst., Leicester, 1994).

Immunomodulation For Research Into Plant Metabolism And Biochemistry. Intracellular expression of antibody molecules can be used to modulate the metabolism of the expressing cell (42,46,47). This approach has numerous potential applications especially as new technologies have demonstrated that antibodies with virtually any specificity can be made (48). In plants, it may be possible to alter plant metabolism by expressing appropriate catalytic antibodies, or enzyme/hormone inactivating antibodies. A good target would be the plant hormones which control growth and development.

This approach has been successfully carried out by the group of Owen et al., who reported the production of transgenic tobacco plants transformed with gene sequences encoding an anti-phytochrome single chain Fv antibody (34). Phytochrome is a family of cytoplasmic proteins which act as photoreceptors and are involved in the photocontrol of tobacco seed germination. Seeds from the transgenic plant line were reported to show aberrant light mediated germination behaviour, as levels of germination were reduced by approximately 40%.

SUMMARY

All heterologous systems for expression of recombinant proteins have advantages and disadvantages. In the case of plants, the most compelling reasons would be the possibility for large scale production, whereby kilogram quantities of antibody can be envisaged, and the ability to produce full length or engineered full length antibodies. For several reasons, antibodies in plants are also of significant interest in plant sciences. The main disadvantages lie in the amount of time required to produce the transgenic antibody producing plant, and the investment in training and equipment required to establish a plant facility.

The ability to glycosylate proteins is of lesser importance in the case of antibodies, where the presence or absence of glycans has little effect on antigen binding, as compared with other proteins where the glycans can be important for the correct folding and function of the protein (49). However, further research is certainly required to investigate the immunogenicity of plant antibodies that are considered for therapeutic applications. Future experiments should also address the mechanisms whereby plant cells are able to assemble antibodies correctly. It is likely that the plant cell employs similar mechanisms to mammalian cells, indeed plants may prove to be a good model in which to study immunoglobulin folding and assembly. Normal cells can be used that produce individual immunoglobulin chains, or assembled combinations, and there is no contamination by other proteins of the immunoglobulin supergene family.

It is also important to determine the characteristics of plants that favour antibody production, in order to make a rational choice of the best plant species to use. Antibody production in plants is still in its infancy, but on the basis of the results achieved so far, plants certainly merit consideration alongside other recombinant expression systems.

LITERATURE CITED.

1. Hiatt, A. C., Cafferkey, R., and Bowdish, K. (1989) *Nature* **342**, 76-78.

2. Horsch, R. B., Fry, J. E., Hoffmann, N. L., Eichholtz, D., Rogers, S. G., and Fraley, R. T. (1985) *Science* **227**, 1229-1231.

3. Rogers, S. G., Horsch, R., and Fraley, R.(1988) in Methods for Plant Molecular Biology (Weissbach, , Ed.), Academic Press, San Diego.

4. Rogers, S. G., Klee, H. J., Horsch, R. B., and Fraley, R. T. (1987) *Meth. Enzymol.* **153**, 253-276.

5. Usha, R., Rohll, J. B., Spall, V. E., Shanks, M., Maule, A. J., Johnson, J. E., and Lomonossoff, G. P. (1993) *Virology* **197**, 366-374.
6. Klein, T. M., Wolff, E. D., Wu, R., and Sanford, J. C. (1987) *Nature* **327**, 70-73.
7. Ma, J. K-C., Lehner, T., Stabila, P., Fux, C. I., and Hiatt, A. (1994) *Eur. J. Immunol.* **24**, 131-138.
8. During, K., Hippe, S., Kreuzaler, F., and Schell, J. (1990) *Plant Mol. Biol.* **15**, 281-293.
9. De Neve, M., De Loose, M., Jacobs, A., Van Houdt, H., Kaluza, B., Weidle, U., and Depicker, A. (1993) *Transgenic Research* **2**, 227-237.
10. Gething, M. J. and Sambrook, J. (1992) *Nature* **355**, 33-45.
11. Melnick, J., Aviel, S., and Argon, Y. (1992) *Journal of Biological Chemistry* **267**, 21303-21306.
12. Hein, M. B., Tang, Y., McLeod, D. A., Janda, K. D., and Hiatt, A. C. (1991) *Biotechnol. Prog.* **7**, 455-461.
13. Fontes, E. B. P., Shank, B. B., Wrobel, R. L., Moose, S. P., O'Brian, G. R., Wurtzel, E. T., and Boston, R. S. (1991) *Plant Cell* **3**, 483-496.
14. Walther-Larsen, H., Brandt, J., Collinge, D. B., and Thordal-Christensen, H. (1993) *Plant Molecular Biology* **21**, 1097-1108.
15. Sturm, A., Van Kuik, J. A., Vliegenthart, J. F., and Chrispeels, M. J. (1987) *Journal of Biological Chemistry* **262**, 13392-13403.
16. Faye, L., Johnson, K. D., Sturm, A., and Chrispeels, M. J. (1989) *Physiol. Plant.* **75**, 309-314.
17. Jones, R. L. and Robinson, D. G. (1989) *New Phytol.* **111**, 567-597.
18. Kornfeld, R. and Kornfeld, S. (1985) *Ann. Rev. Biochem.* **54**, 631-664.
19. Kimura, Y., Hase, S., Kobayashi, Y., Kyogoku, Y., Ikenaka, T., and Funatsu, G. (1988) *Journal of Biochemistry* **103**, 944-949.
20. Sturm, A. (1991) *Eur. J. Biochem.* **199**, 169-179.
21. von Schaewen, A., Sturm, A., O'Neill, J., and Chrispeels, M. J. (1993) *Plant Physiol.* **102**, 1109-1118.
22. Carpita, N., Sabularse, D., Montezinos, D., and Helmer, D. P. (1979) *Science* **205**, 1144-1147.
23. Firek, S., Draper, J., Owen, M. R. L., Gandecha, A., Cockburn, B., and Whitelam, G. C. (1993) *Plant Molecular Biology* **23**, 861-870.
24. Vierling, E. (1991) *Annu. Rev. Plant Physiol. Plant Mol. Biol.* **42**, 579-620.
25. Roy, H. (1989) *Plant Cell* **1**, 1035-1042.
26. Nagata, T., Okada, K., Kawazu, T., and Tabeke, I. (1987) *Mol. Gen. Genet.* **207**, 242-244.
27. Williamson, J. D., Hirsch, M. E., Larkins, B. A., and Gelvin, S. B. (1989) *Plant Physiol.* **90**, 1570-1576.
28. Better, M., Chang, C. P., Robinson, R. R., and Horwitz, A. H. (1988) *Science;* **240**, 1041-1043.
29. Skerra, A. and Pluckthun, A. (1988) *Science* **240**, 1038-1040.
30. Huston, J. S., Levinson, D., Mudgett-Hunter, M., Tai, M. S., Novotny, J., Margolies, M. N., Ridge, R. J., Bruccoleri, R. E., Haber, E., Crea, R., and Oppermann, H. (1988) *Proc. Natl. Acad. Sci. USA.* **85**, 5879-5883.
31. Bird, R. E., Hardman, K. D., Jacobson, J. W., Johnson, S., Kaufman, B. M., Lee, S. M., Lee, T., Pope, S. H., Riordan, G. S., and Whitlow, M. (1988) *Science* **242**, 423-426.
32. Ward, E. S., Gussow, D., Griffiths, A. D., Jones, P. T., and Winter, G. (1989) *Nature;* **341**, 544-546.
33. Benvenuto, E., Ordas, R. J., Tavazza, R., Ancora, G., Biocca, S., Cattaneo, A., and Galeffi, P. (1991) *Plant Mol. Biol.* **17**, 865-874.
34. Owen, M., Gandecha, A., Cockburn, B., and Whitelam, G. (1992) *Biotech.* **10**, 790-794.
35. Tavladoraki, P., Benvenuto, E., Trinca, S., De Martinis, D., Cattaneo, A., and Galeffi, P. (1993) *Nature* **366**, 469-472.
36. Ma, J. K-C., Hunjan, M., Smith, R., Kelly, C., and Lehner, T. (1990) *Infect. Immun.* **58**, 3407-3414.
37. Hiatt, A. (1990) *Nature* **344**, 469-470.

38. Lehner, T., Caldwell, J., and Smith, R. (1985) *Infect. Immun.* **50**, 796-799.
39. Michalek, S. M., Gregory, R. L., Harmon, C. C., Katz, J., Richardson, G. J., Hilton, T., Filler, S. J., and McGhee, J. R. (1987) *Infect. Immun.* **55**, 2341-2347.
40. Filler, S. J., Gregory, R. L., Michalek, S. M., Katz, J., and McGhee, J. R. (1991) *Archs. Oral Biol.* **36**, 41-47.
41. Hamada, S., Horikoshi, T., Minami, T., Kawabata, S., Hiraoka, J., Fujiwara, T., and Ooshima, T. (1991) *Infect. Immun.* **59**, 4161-4167.
42. Werge, T. M., Biocca, S., and Cattaneo, A. (1990) *FEBS Lett.* **274**, 193-198.
43. Piccioli, P., Ruberti, F., Biocca, S., Di Luzio, A., Werge, T. M., Bradbury, A., and Cattaneo, A. (1991) *Proc. Natl. Acad. Sci. USA.* **88**, 5611-5615.
44. Dyer, M. J., Hale, G., Hayhoe, F. G., and Waldmann, H. (1989) *Blood.* **73**, 1431-1439.
45. Caron, P. C., Laird, W., Co, M. S., Avdalovic, N. M., Queen, C., and Scheinberg, D. A. (1992) *J. exp. Med.* **176**, 1191-1195.
46. Tang, Y., Hicks, J. B., and Hilvert, D. (1991) *Proceedings of the National Academy of Sciences of the United States of America* **88**, 8784-8786.
47. Carlson, J. R. (1988) *Molecular & Cellular Biology* **8**, 2638-2646.
48. Nissim, A., Hoogenboom, H. R., Tomlinson, I. M., Flynn, G., Midgley, C., Lane, D., and Winter, G. (1994) *EMBO Journal* **13**, 692-698.
49. Li, Y., Luo, L., Rasool, N., and Kang, C. Y. (1993) *Journal of Virology* **67**, 584-588.

RECEIVED May 17, 1995

Chapter 6

Engineering of Immunoglobulin Fc and Single-Chain Fv Proteins in *Escherichia coli*

David Filpula[1], Michele Rollence[1], Nina Essig[1], James Nagle[2], Aniruddha Achari[2], and Timothy Lee[1]

[1]Enzon, Inc., 20 Kingsbridge Road, Piscataway, NJ 08854
[2]Genex Corp., 16020 Industrial Drive, Gaithersburg, MD 20877

The field of antibody engineering encompasses the investigation and redesign of constant domain associated effector functions, as well as the intensive current research on variable domain associated antigen-binding sites. We have investigated the potential of *E. coli* expression of single-chain Fv and Fc proteins to both provide model systems and practical reagents for antibody effector region interactions. We shall first review the current status of single-chain Fv protein engineering. A selected single-chain Fv protein with human IgM C_μ specificity has been constructed and characterized in our laboratory. The variable region genes of anti-μ monoclonal antibody DA4.4 were isolated and an engineered sFv version of this antibody expressed in *E. coli* was shown to be an IgM-specific binding protein. A simple screening method was developed for protein-protein interactions between recombinant C_H-producing *E. coli* and C_H-binding proteins displayed on the surface of bacteriophage M13. This method, which allows facile engineering and recovery of recombinant gene products, provides a model system for the characterization and designed modification of antibody effector domains.

The development of monoclonal antibody (MAb) technology in 1975 allowed the establishment and maintenance of hybridoma cell lines which secrete antibodies of a unique chosen specificity (*1*). The potential value of MAbs with defined specificities as imaging or therapeutic agents is apparent (*2*). However, further genetic engineering of affinity, isotype and effector functions may be required to produce antibody proteins with the desired biological activity *in vivo*. The recent extension of MAb technology to the expression of cloned antibody fragments in microbial hosts has made antibody engineering a

0097–6156/95/0604–0070$12.00/0

flourishing area of research (*3*). The current focus of antibody engineering in *E. coli* has been the expression of antigen-binding fragments which may lack part or all of the antibody constant domains as shown in Figure 1. Antibody effector functions, such as Fc receptor targeting and complement fixation are derived from the C_H domains. The Fv fragment is a two-chain heterodimer of the antibody variable domains (V_L and V_H) and is regarded as the minimal antigen-binding fragment. At low protein concentrations, Fv have been found to dissociate into V_L and V_H domains (*4*). This limited stability may be substantially overcome by incorporation of a designed linker peptide to bridge isolated V_L and V_H domains into a single polypeptide (*5,6*) as shown in Figure 2. Basic research on recombinant single-chain Fv (sFv) proteins (often designated single-chain antigen-binding proteins or SCA) has provided an abundance of information about the value of this new antibody technology.

sFv Structural and Functional Integrity. A number of antigen-binding studies reported for sFv proteins recognizing antigens derived from haptens, polypeptides, carbohydrates and cell surfaces indicate that the binding site fidelity and the affinity constants of sFv proteins may be equivalent to the MAbs from which they were derived (*7,8*). Independent evidence for the identity of the binding site in MAbs and sFv proteins is provided by the demonstration of equivalent kinetic parameters of a catalytic MAb and its derived sFv version (*9*). NMR spectroscopy (*10*) and X-ray crystallography (*11,12*) also support the structural fidelity of the sFv binding site.

Anti-tumor sFv. A major focus of our work on sFv proteins has been the development of anti-tumor immunotargeting agents. When compared to MAbs *in vivo*, anti-tumor sFv proteins display rapid blood clearance and penetrate tumors swiftly and with even distribution (*13,14,15*). We have developed an sFv derived from MAb CC49 and compared the *in vivo* pharmacokinetic properties with intact antibody, Fab' and F(ab')$_2$. MAb CC49 is a murine IgG that reacts to the pancarcinoma antigen TAG-72 (*16*). Clinical trials using ^{131}I-labeled CC49 have demonstrated efficient tumor targeting of colorectal carcinoma lesions (*17,18*). The ^{131}I-labeled sFv was shown to bind biopsies of TAG-72 expressing tumors. Metabolism studies in mice demonstrated an extremely rapid plasma and whole body clearance for the sFv as compared to intact IgG or antibody fragments Fab' and F(ab')$_2$. The relative K_a values of dimeric intact IgG and F(ab')$_2$ were shown to be about eightfold higher than those of the monomeric Fab' and sFv, which is presumably a consequence of the difference in binding avidity to the repetitive carbohydrate epitopes of TAG-72. In biodistribution studies, using the LS-174T human colon carcinoma xenograft in athymic mice, the percentage injected dose/g in the tumor at 24 h for monomeric sFv and Fab' were about tenfold lower than for the dimeric IgG and F(ab')$_2$. However, tumor : normal or signal : noise ratios (radiolocalization indices) for the sFv were equivalent or greater than those for the other antibody forms, although at different time intervals (*13*). The high kidney uptake in both F(ab')$_2$ and Fab', which have C_L and C_H1 domains, was

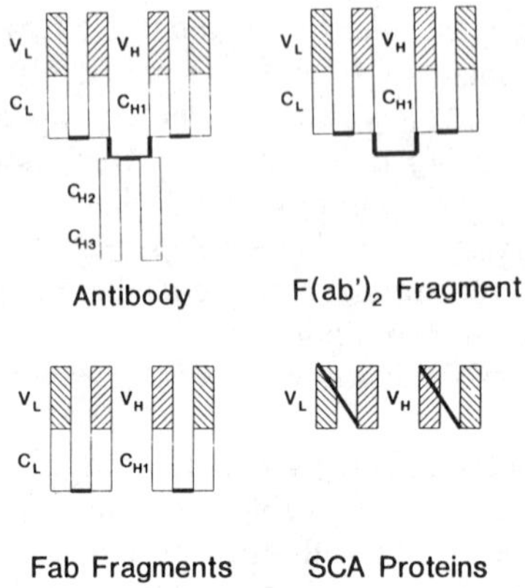

Figure 1. Schematic diagram of IgG antibody and three antibody fragments. Thatched rectangles contain antigen-binding sites (Fv) derived from variable light chain (V_L) and variable heavy chain (V_H) domains. Open rectangles contain constant heavy chain (C_H) or light chain (C_L) domains. Bold horizontal lines denote interchain disulfides. Bold vertical lines denote hinge region bridging C_{H1} and C_{H2}.

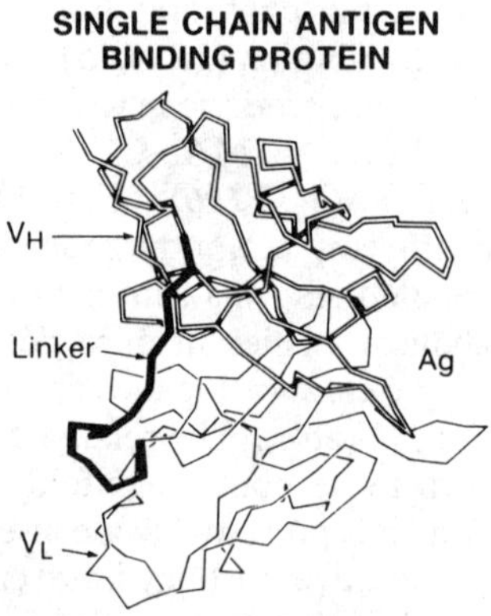

Figure 2. Structural view of a single-chain Fv fragment. The antigen-binding site (Ag) and the linker sequence are indicated.

not observed with CC49 sFv. Quantitative autoradiography to define tumor penetration of the antibody species revealed that most of the intact IgG delivered to the tumor was concentrated in the region adjacent to blood vessels, while the sFv was rapidly and evenly distributed throughout the tumor mass (*14*). The Fab' and F(ab')$_2$ distributions showed intermediate penetration. The intact IgG reached a degree of penetration at 48 to 96 h postinjection equivalent to the sFv at 0.5 h postinjection. This extremely rapid tumor penetration can be significant in cancer therapy applications and suggests that tumor tissue barriers may be size dependent. CC49 sFv is currently under development for clinical applications in the management of cancer. While rapid clearance of the sFv from blood may limit toxicity, the rapid and homogeneous tumor penetration may offer a more successful delivery of the sFv or an sFv-toxin.

An important application of sFv proteins will be in the production of recombinant single-chain immunoeffector proteins which link an sFv-derived cell binding specificity to an effector protein domain. Single-chain immunotoxins derived from anti-tumor Fv and *Pseudomonas* exotoxin A have been shown to regress established tumors in rats and mice (*19,20*). In addition, the expression of antibody specificities in the form of a single transcript, single subunit sFv may be valuable in gene therapy applications of targeted viral vectors (*21*) or intracellular antibodies (*22*).

Multivalent Fv. In natural antibodies, the V_L and V_H domains associate through noncovalent interactions. The designed linker peptide in sFv proteins creates a single subunit protein with two variable domains. sFv proteins lack the constant domains and bivalency of MAbs. MAbs (150 kDa) are sixfold larger than sFv proteins and display high avidity for repetitive epitopes. However, recent reports describe an approach to the development of multivalent Fv from minimal sFv proteins (*23,24,25,26*). Single-chain Fv proteins are known to aggregate and form multimeric species. These stable Fv dimers and higher aggregates result from intermolecular V_L/V_H pairings between two or more sFv polypeptides to produce multivalent Fv molecules from monovalent sFv as shown in Figure 3. The extent of multimeric Fv formation is influenced by linker length as well as the specific variable domain sequences. In general, shorter linkers have an increased aptitude to form dimers and larger aggregates. An sFv protein with a short linker (0-10 residues), which restrains intramolecular V_L/V_H domain pairings, forms predominately multimeric Fv (*23*). The discovery of rearranged multivalent Fv may significantly extend sFv technology into bispecificity and cross-linking capabilities without necessarily increasing the minimal sFv polypeptide chain length. For example, we produced a functional heterodimer from the CC49 sFv and anti-fluorescein sFv 4-4-20 by variable domain shuffling (*24*). The first mixed sFv protein had a 4-4-20 V_L-linker-CC49 V_H construction. The second mixed sFv construction had the CC49 V_L-linker-4-4-20 V_H sequence. The resulting 4-4-20/CC49 two-chain heterodimer bound both fluorescein and TAG-72 antigen. The stability in solution of multivalent Fv is currently under investigation (24,25,26).

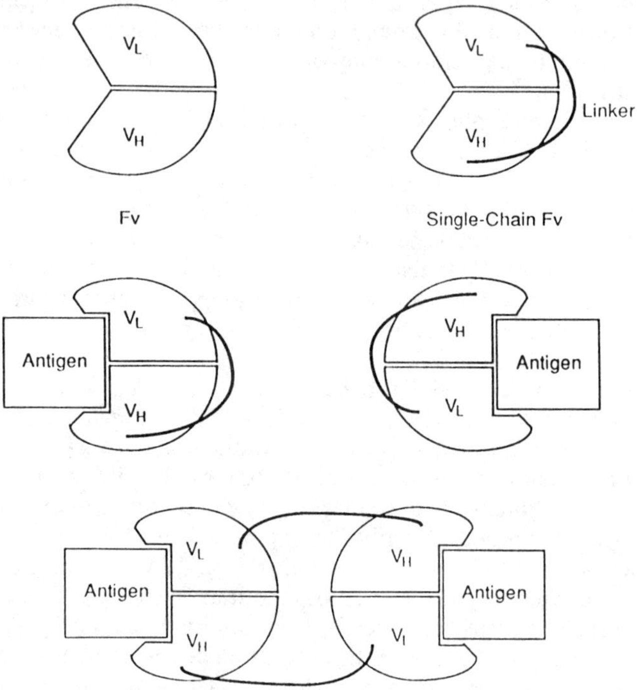

Figure 3. Schematic diagram of monomeric and multimeric Fv. Top row: Fv and sFv. Middle row: sFv bound to antigen. Bottom row: Fv dimer formed from intermolecular V_L/V_H pairings.

Variable Domain Orientation and Linker Design. Since initial sFv linkers were designed to covalently connect the two variable domains without participation in the final Fv conformation, it might be predicted that active sFv proteins can be constructed in either domain orientation. Either V_L is the N-terminal domain followed by the linker and V_H (V_L-linker-V_H construction) or V_H is the N-terminal domain followed by the linker and V_L (V_H-linker-V_L construction). Both types of purified recombinant sFv proteins have been reported to retain MAb binding site specificities and affinities (*7,8*). The choice of variable domain orientation can be relevant to optimal retention of binding activity however, and individual Fv binding sites and selected linkers may differ in this characteristic. For example, Desplanq, et al (*27*) found much greater binding activity with a V_L-linker-V_H construction of B72.3 sFv. Anand, et al (*28*) investigated the effects of variable domain order on Se155-4 sFv and found that the V_H-linker-V_L construction was tenfold higher in affinity than the V_L-linker-V_H. From these and other studies (*25,29*), it appears that the optimal variable domain orientation is dependent on specific Fv, and binding activity may often be similar in either orientation.

The polypeptide linker sequences in sFv proteins were designed to span the 3.5-4.0 nm between the C-terminus of V_L (or V_H) and the N-terminus of V_H (or V_L). To furnish flexibility, the linkers commonly include a motif of alternating glycine and serine residues. This rationale is also followed in the $(G_4S)_n$ linkers described by Huston, et al (*6*) and serine-rich linkers described by Dorai, et al (*30*). Our current linker designs (*24,31*) of 12 to 18 residues include also three charged residues to further enhance the solubility of the linker and its associated sFv, without increasing the susceptibility of the linker to proteolysis.

Engineering of C_H-Binding Proteins. During the development of sFv methods in our anti-tumor antibody research, we recognized an opportunity and a utility to applying these protein engineering and microbial expression systems to the investigation of the other segment of natural antibodies corresponding to the constant domains which specify cellular effector activities. Since bacteria lack glycosylation mechanisms, recombinant IgG Fc proteins from *E. coli* will be deficient in some effector functions which require C_H2 associated oligosaccharide attachment (*32*). However, unglycosylated Fc has been reported to retain Clq and protein A binding capacity (*33*). The production of recombinant Fc of other antibody isotypes (IgM, IgA, IgE, IgD) would also provide a new experimental approach to investigating the glycosylation dependence of associated effector functions, such as cell surface Fc receptor binding and complement fixation. Furthermore, it would be beneficial to develop C_H-specific binding proteins which recognize polypeptide sequence-specific epitopes of immunoglobulins for use as both research tools and as clinical reagents. Using the same expression systems we employed for sFv production, we developed model systems for C_H-binding proteins. First, active recombinant IgG1 Fc was expressed in *E. coli.* Second, a novel screening method was developed for protein interactions between recombinant Fc and a recombinant Fc-binding domain. Third, an sFv protein with C_μ specificity derived from an anti-IgM MAb was constructed and characterized.

Recombinant Fc of Human IgG1 from *E. coli.* The Fc fragment of IgG consists of two identical polypeptide chains from the hinge-C_H2-C_H3 segment which are linked by interchain hinge disulfides as shown in Figure 1. In terms of molecular weight and immunoglobulin domain structure, one Fc fragment chain and an sFv protein are physically similar. We employed our standard sFv expression system for Fc production. The vector for sFv expression in *E. coli* contains the hybrid lambda phage promoter O_L/P_R and the ompA signal sequence as shown in Figure 4. To produce the final expression strains, the completed sFv expression vectors are transformed into *E. coli* host strain GX6712 which contains the gene for the cI*ts*857 temperature-sensitive repressor. This provides a transcriptional regulation system in which induction of sFv synthesis occurs by raising the culture temperature from 32°C to 42°C. About twenty independent rodent-derived sFv proteins have been expressed at 5-20% of total cell protein using this expression system (*7,25*). N-terminal amino acid sequence analysis has confirmed the predicted signal sequence removal. Inclusion of the signal sequence in our standard expression system has provided mature sFv production at high levels for several distinct sFv proteins, although overexpression of the secreted proteins in *E. coli* does result in the formation of protein aggregates in the periplasmic space, as well as partial lysis. The insoluble aggregates are subjected to denaturation and refolding prior to purification.

The purification protocol for sFv involves resuspension in guanidine denaturing solvent, refolding by dilution and cation exchange HPLC steps (*7,25*). Per liter of fermentation volume, this protocol yields 5-20 milligram amounts of sFv proteins that are more than 95% pure as examined by SDS PAGE and Scatchard analysis.

The Fc polypeptide was produced at approximately 10% of total cell protein by this expression system. Following our standard resolubilization and refolding conditions (*7,25*) the Fc protein was purified by a single affinity chromatography step on Protein G agarose. The binding and elution patterns of the recombinant Fc were similar to native IgG purified on the same column. Our characterization of the purified recombinant Fc protein showed that at least two native Fc functions, protein G affinity and protein A affinity, are fully retained in agreement with a previous report (*33*). However, SDS PAGE in the absence of 2-mercaptoethanol indicated a lack of interchain disulfide formation in the hinge region, under our refolding and purification conditions. This result suggests that native Fc conformation can form without hinge covalent linkage. The ordered dimeric structure from a "single-chain Fc" polypeptide pair contrasts to Fv dimer formation from single-chain Fv where alternate intramolecular and intermolecular domain pairings are possible (Figure 3). The crystallization of recombinant Fc complexes for structural studies is under investigation.

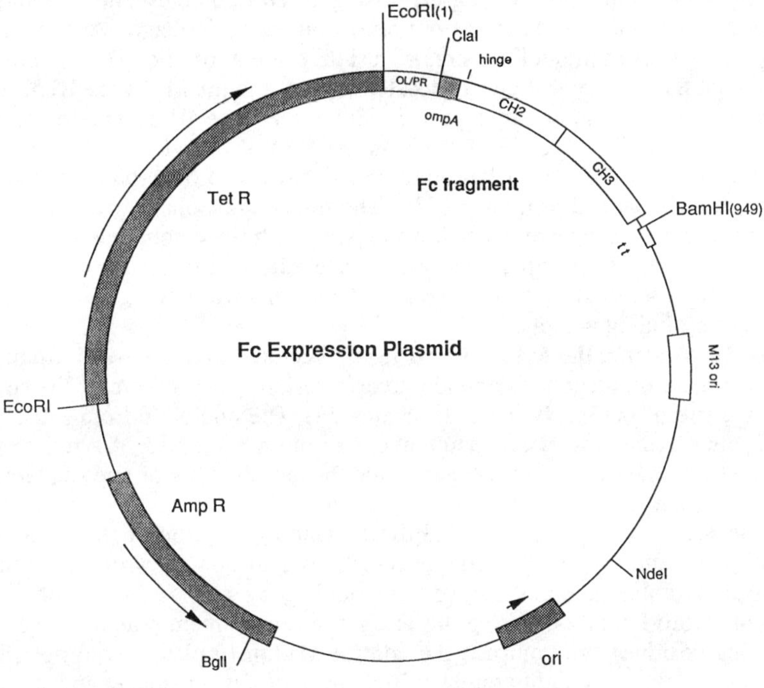


```
                5              10             15             20             25             30
  1 M K K T A I A I A V A L A G F A T V A Q A D V E S K S C D K
 31 T H T C P P C P A P E L L G G P S V F L F P P K P K D T L M
 61 I S R T P E V T C V V V D V S H E D P E V K F N W Y V D G V
 91 E V H N A K T K P R E E Q Y N S T Y R V V S V L T V L H Q D
121 W L N G K E Y K C K V S N K A L P A P I E K T I S K A K G Q
151 P R E P Q V Y T L P P S R D E L T K N Q V S L T C L V K G F
181 Y P S D I A V E W E S N G Q P E N N Y K T T P P V L D S D G
211 S F F L Y S K L T V D K S R W Q Q G N V F S C S V M H E A L
241 H N H Y T Q K S L S L S P G K *
```

Figure 4. Expression vector used for Fc (and sFv) production in *E. coli.* Top: Plasmid contains the hybrid lambda phage promoter O_L/P_R, the ompA signal sequence and a hinge-C_H2-C_H3 gene. Bottom: Translation product with mature protein N-terminus at residue 22.

Screening Method for Interacting Recombinant Proteins. The display of antibody antigen-binding fragments on the surface of filamentous phage has been utilized for enrichment of selected antigen specificities and the engineering of binding site affinity (*34*). We wanted to develop a simple filter screening method for the direct identification and subsequent engineering of protein interactions between C_H-expressing *E. coli* colonies and C_H-binding proteins expressed on the surface of filamentous bacteriophage. As the phage component of this model system, a fusion protein of the 58 residue FB (Fragment B) domain of staphylococcal protein A and the M13 gene III protein was constructed as shown in Figure 5. Since the gene III surface protein is present in about 3 to 5 copies per virion, an M13 phage with a monomer FB domain fused to the gene III protein would have 3 to 5 Fc-binding domain copies as does natural protein A (*35*). The filamentous phage group includes M13 as well as fd phage and we chose to work with the completely sequenced M13mp19 (*36*). Prior experience with phage surface expression of the anti-fluorescein 4-4-20 sFv in this system had demonstrated a retention of fluorescein binding activity.

To construct the M13 gene III fusion vector, a *Sty*I site was introduced by site-directed-mutagenesis one amino acid residue downstream of the signal cleavage site of M13mp19 gene III (Figure 5). The non-palindromic *Sty*I site allows obligate head-to-tail insertion of one or more FB copies. A synthetic FB domain with *Sty*I overhangs was ligated into the modified vector and single-copy FB recombinants were identified by DNA sequencing. In order to demonstrate that the M13 fusion phage retained the Fc-binding function of the protein A domains, plaques were lifted on nitrocellulose, incubated with peroxidase-conjugated rabbit IgG and assayed for binding by color development with H_2O_2/4-chloro-l-napthol. Using this assay, the recombinant plaques appeared as strong positives. In contrast, the control M13mp19 plaques did not show detectable color. A deliberately mixed lawn of fusion phage and control M13mp19 was also plated and processed through this filter assay. After identification of positive and negative plaques, DNA sequence analysis confirmed that the IgG-binding plaques have an FB domain in gene III, whereas the negative plaques did not. Similar results were reported recently for M13 phage with a single FB domain per virion (*37*). Soluble protein A or polyclonal human IgG were demonstrated to exhibit competitive inhibition in the binding assay. When the protein A fusion phage were screened with peroxidase-conjugated goat IgG, a very weak signal was observed in agreement with the weak affinity of natural protein A for goat IgG.

E. coli containing the expression vector for the human IgG1 Fc fragment were streaked out on LB/ampicillin agar at 30°C to obtain small colonies which were lifted onto a nitrocellulose filter. The filter was incubated on a fresh LB agar plate for 6 hours at 42°C to induce Fc expression, washed with TBST buffer (*38*), and BSA blocking solution (*38*) and then cut in half. One half of the filter was incubated in BSA blocking solution plus FB fusion phage

(10^{10} phage per ml); the other half was incubated in BSA blocking solution plus M13mp19 phage (10^{10} phage per ml). After a six hour incubation at 22°C, the filters were washed with TBST four times at 5 minutes per wash; then washed once with TBS for 5 minutes and dried in a vacuum oven at 80°C for 2 hours. The dried filters were incubated in pre-hybridization buffer (*39*) for 2 hours at 47°C and then incubated with ^{32}P-labeled M13-specific oligonucleotide probes for 2 hours at 47°C. After three washes with 2 X SSC (*39*) at 47°C, the filters were exposed to X-ray film with one intensifying screen. As shown in the top of Figure 6, the filter-half incubated with FB fusion phage showed distinct phage binding to the bacterially expressed Fc and the X-ray film exposure corresponded to the colony pattern on the agar plate. The filter-half incubated with the control M13mp19 showed no response. When the Fc expression strain was grown at 30°C (and hence uninduced for Fc expression), neither FB fusion phage nor M13mp19 incubation resulted in a detectable signal, as shown in the bottom of Figure 6. These results, done in triplicate, demonstrate the development of a simple screening method for identifying protein-protein interactions between two cloned gene products. We have previously demonstrated that the Fc expression system results in spontaneous cell lysis (*7,25*) and hence the expressed proteins may be directly bound by the nitrocellulose filters. Competitive inhibition by soluble protein A and polyclonal IgG are further indications of a specific protein A-Fc interaction in this method. A simple modification of this procedure using *in vitro* protein labeling of the phage coat could allow direct detection of filter-bound phage and would also facilitate subsequent recovery of viable probe-phage bound to the filter. The Fc-producing bacterial colonies can conveniently be recovered from the master plate. The proportion of Fc protein bound by the nitrocellulose which is correctly folded is unknown, but is sufficient for easy detection by this method. Because both interacting cloned gene products may be selectively or randomly mutated and subsequently recovered after screening, we are employing this system in the investigation and engineering of targeted C_H- binding proteins, including anti-C_H sFv proteins.

Construction and Characterization of a C_μ-Binding sFv. We selected the μ-chain of IgM as our initial target for development of a C_H-binding sFv molecule. IgM is produced as a first response to antigenic challenge in mammals and accounts for about 10% of total human immunoglobulins (*38*). Pathological IgM accumulates in blood in certain disease states such as myeloma and autoimmune disease. Many antibodies of the IgM isotype are presently being excluded from research since it is difficult to recover them in a pure and active form. The IgG-binding staphylococcal protein A has been one of the most widely used reagents in antibody research. The value of developing sequence-specific single-subunit binding molecules for IgM and other antibody isotypes is apparent.

The MAb DA4.4 (ATCC HB57) described by Maruyama, et al (*40*) is a murine IgG1*k* antibody which is specific for human μ-chains. The ATCC antibody was confirmed to be specific for μ-chains by Western blot analysis.

```
                                                 /
  K  K  L  L  F  A  I  P  L  V  V  P  F  Y  S  H  S  A  A  K  A  D  N  K  F  N  K  E  Q
GTGAAAAAATTATTATTCGCAATTCCTTTAGTTGTTCCTTTCTATTCTCACTCCGCTGCCAAGGCTGACAACAAATTCAACAAAGAACAG
                                                          StyI

 Q  N  A  F  Y  E  I  L  H  L  P  N  L  N  E  E  Q  R  N  G  F  I  Q  S  L  K  D  D  P  S
CAGAACGCTTTCTACGAAATCCTGCACCTGCCGAACCTGAACGAAGAACAGCGTAACGGTTTCATCCAGTCTCTGAAAGACGACCCGTCT

 Q  S  A  N  L  L  A  E  A  K  K  L  N  D  A  Q  A  P  K  V  E  S  C  L  A  K  S  H
CAGTCTGCTAACCTGCTGGCTGAAGCTAAAAAACTGAACGACGCTCAGGCACCCAAGGTTGAAAGTTGTTTAGCAAAATCCCAT------
                                                    StyI
```

Figure 5. DNA sequence and translation product of the fusion of protein A domain (underlined) to M13 gene III protein. The signal cleavage site is indicated by /.

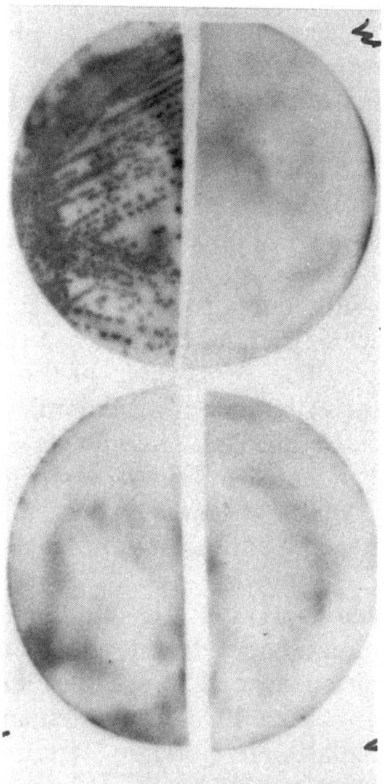

Figure 6. Filter screening method for interacting recombinant proteins. Fc protein from induced bacteria is bound to nitrocellulose filters which are screened with phage probes. Upper left: Bacterial Fc expression is induced; probe is protein A phage. Upper right: Bacterial Fc expression in induced; probe is M13mp19 phage. Lower left: Bacterial Fc expression is uninduced; probe is protein A phage. Lower right: Bacterial Fc expression is uninduced; probe is M13mp19 phage.

Dot blot analysis showed a binding specificity for IgM, but not IgG. Following cDNA cloning of the V_L and V_H genes from DA4.4 hybridoma cells, a single-chain Fv protein version of this antibody was constructed (*41*). The fourteen amino acid linker "212" has been previously utilized and was chosen for the V_L-linker-V_H construction shown in Figure 7. As with our previously tested sFv proteins, the DA4.4 sFv was expressed as insoluble protein at about 10% of

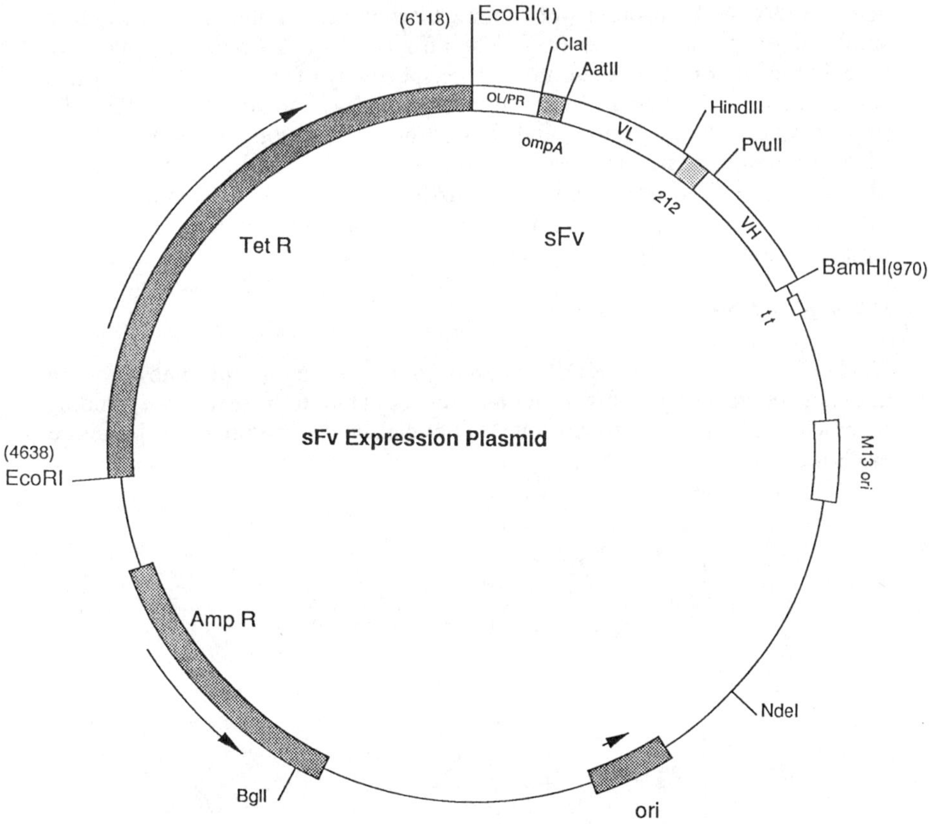

DVVMTQSPSSLAMSVGQKVTMSC**KSSQSLLNSSNQKNYLA**WYQQKPGQSPELLVY**FASTRES**GVP

DRFIGSGSGTDFTLTISSVQAEDLADYFC**QQHYSTPFT**FGSGTKLEIKGSTSGSGKSSEGKGQVQ

LQQPGAEFVKPGAPVKLSCKASGYPFT**TYWVN**WMKQRPGRGLEWIG**RIDPYDSETLYNQKFKD**KA

TLTVDKSSSTAYIQLSSLTSEDSAVYYCAR**ETYDYPFAY**WGQGTLVTVS

Figure 7. Expression vector used for sFv production in *E. coli*. Top: Plasmid contains O_L/P_R promoter, ompA signal sequence and a V_L-linker-V_H sFv gene. Bottom: Amino acid sequence of mature DA4.4/212 sFv protein with 212 linker underlined and CDR regions in bold print.

total cell protein. SDS PAGE analysis indicated that the sFv protein exhibited the expected molecular weight of 26,812. The detailed refolding and purification protocol of the DA4.4 sFv protein has been described (*7,41*). Competitive ELISA was used to show that the DA4.4 sFv is an IgM-binding protein. The DA4.4 MAb was biotinylated to act as a tracer molecule with which either the unlabeled MAb or the sFv protein would compete. Biotin modification of the MAb did not affect IgM-binding. Both the DA4.4 MAb and the DA4.4 sFv protein were found to compete off the labeled MAb in similar dose response curves (*41*). The affinity of the MAb was estimated to be 2.5 times that of the sFv protein. The specificity of the DA4.4 sFv protein was further established in an experiment in which an anti-fluorescein sFv protein was used as a control and showed no competition against the DA4.4 MAb for binding to human IgM as shown in Figure 8. The DA4.4 sFv protein, on the other hand, competes with its MAb counterpart in a dose dependent manner. In an analogous series of experiments, we have also developed an sFv with anti-mouse IgM specificity (*41*).

CONCLUSIONS

Single-chain Fv technology has proven to be an important innovation in antibody engineering. sFv proteins may be able to provide the binding specificity and affinity of monoclonal antibodies *in vivo*, but in a smaller, more easily modified molecule. Significantly, pre-clinical research studies have shown

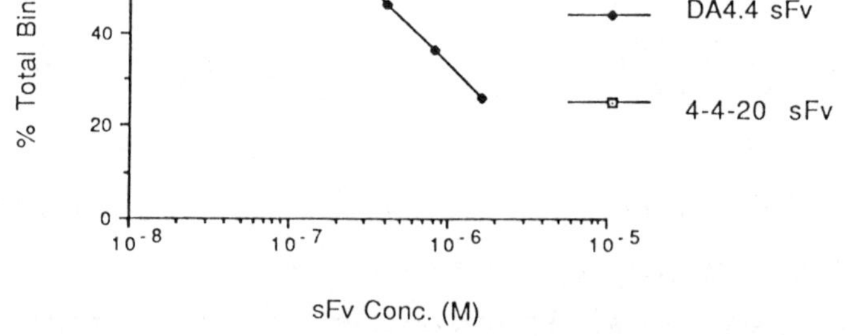

Figure 8. ELISA showing competition between the DA4.4 MAb against its sFv counterpart for immobilized human IgM. A control anti-fluorescein sFv 4-4-20 showed no competition against the DA4.4 MAb for binding to human IgM. The signal in this ELISA is generated by a goat anti-mouse IgG conjugated to HRP, instead of streptavidin-HRP.

that anti-tumor sFv proteins rapidly target and penetrate tumors with a homogeneous distribution.

The *E. coli* genetic expression systems developed in sFv research may also be applied to the investigation and engineering of the constant domains of antibodies. The engineering of specific C_H-binding proteins is a practical application of sFv technology under development.

ACKNOWLEDGEMENTS

This work was supported in part by NIH grant #1R43AI30747-01, awarded under the Small Business Innovation Research Program. We thank our colleagues at Enzon, Genex and NCI for their contributions to this research program.

REFERENCES

1. Köhler, G.; Milstein, C. *Nature* **1975**, *245*, 495-497.
2. Waldmann, T. A. *Science* **1991**, *252*, 1657-1662.
3. Winter, G.; Milstein, C. *Nature* **1991**, *349*, 293-299.
4. Glockshuber, R.; Malia, M.; Pfitzinger, I.; Plückthun, A. *Biochemistry* **1990**, *29*, 1362-1367.
5. Bird, R. E.; Hardman, K. D.; Jacobson, J. W.; Johnson, S.; Kaufman, B. M.; Lee, S.-M.; Lee, T.; Pope, S. H.; Riordan, G. S.; Whitlow, M. *Science* **1988**, *242*, 423-426.
6. Huston, J. S.; Levinson, D.; Mudgett-Hunter, M.; Tai, M.-S.; Novotny, J.; Margolies, M. N.; Ridge, R. J.; Bruccoleri, R. E.; Haber, E.; Crea, R.; Oppermann, H. *Proc. Natl. Acad. Sci.,* USA **1988**, *85*, 5879-5883.
7. Whitlow, M.; Filpula, D. *Methods, Companion Methods Enzymol* **1991**, *2*, 97-105.
8. Huston, J. S.; Mudgett-Hunter, M.; Tai, M.-S.; McCartney, J.; Warren, F.; Haber, E.; Oppermann, H. In *Methods in Enzymology*; Langone, J. J., Ed.; Academic Press: NY, **1991**, vol. 203; pp. 46-88.
9. Gibbs, R. A.; Posner, B. A.; Filpula, D. R.; Dodd, S. W.; Finkelman, M. A. J.; Lee, T. K.; Wroble, M.; Whitlow, M.; Benkovic, S. J. *Proc. Natl. Acad. Sci., USA* **1991**, *88*, 4001-4004.
10. Freund, C.; Ross, A.; Plückthun A.; Holak. T. A. *Biochemistry* **1994**, *33*, 3296-3303.
11. Zdanov, A.; Li, Y.; Bundle, D. R.; Deng, S.-J.; MacKenzie, C. R.; Narang, S. A.; Young, N. M.; Cygler, M. *Proc. Natl. Acad. Sci., USA* **1994**, *91*, 6423-6427.
12. Kortt, A. A.; Malby, R. L.; Caldwell, J. B.; Gruen, L. C.; Ivancic, N.; Lawrence, M. C.; Howlett, G. J.; Webster, R. G.; Hudson, P. J.; Colman, P. M. *Eur. J. Biochem* **1994**, *221*, 151-157.
13. Milenic, D. E.; Yokota, T.; Filpula, D. R.; Finkelman, M. A. J.; Dodd, S. W.; Wood, J. F.; Whitlow, M.; Snoy, P.; Schlom, J. *Cancer Res.* **1991**, *51*, 6363-6371.

14. Yokota, T.; Milenic, D. E.; Whitlow, M.; Schlom, J. *Cancer Res.* **1992,** *52*, 3402-3408.
15. Whitlow, M.; Filpula, D. In *Tumor Immunobiology: A Practical Approach*; Gallagher, G.; Rees, R. C.; Reynolds, C. W., Eds.; IRL Press: Oxford, England, **1993**, pp. 279-291.
16. Muraro, R.; Kuroki, M.; Wunderlich, D.; Poole, D. J.; Colcher, D.; Thor, A.; Greiner, J. W.; Simpson, J. F.; Molinolo, A.; Noguchi, P.; Schlom, J. *Cancer Res.* **1988**, *48*, 4588-4596.
17. Divgi, C. R.; Kemeny, N.; Kelsen, D.; Cordon-Cardo, C.; Finn, R. D.; Schlom, J.; Houghton, A. N.; Larson, S. M. *Antibody Immunoconj. Radiophar.* **1992,** *5*, 135.
18. Houchens, D.; Cote, R.; Saad, A.; Nines, R.; Berens, A.; Crockford, D.; Rojas-Corona, R.; Schneebaum, S.; Arnold, M.; Martin, E., Jr. *Antibody Immunoconj. Radiophar.* **1992**, *5*, 135.
19. Brinkmann, U.; Pai, L. H.; FitzGerald, D. J.; Willingham, M. C.; Pastan, I. *Proc. Natl. Acad. Sci., USA* **1991**, *88*, 8616-8620.
20. Siegall, C. B.; Chace, D.; Mixan, B.; Garrigues, U.; Wan, H.; Paul, L.; Wolff, E.; Hellström, I.; Hellström, K. E. *J. Immunol.* **1994**, *152*, 2377-2384.
21. Russel, S. J.; Hawkins, R. E.; Winter, G. *Nucleic Acids Res.* **1993,** *21*, 1081-1085.
22. Marasco, W. A.; Haseltine, W. A.; Chen, S. *Proc. Natl. Acad. Sci., USA* **1993**, *90*, 7889-7893.
23. Holliger, P.; Prospero, T.; Winter, G. *Proc. Natl. Acad. Sci., USA* **1993,** *90*, 6444-6448.
24. Whitlow, M.; Filpula, D.; Rollence, M. L.; Feng, S.-L.; Wood, J. F. *Protein Engng.* **1994**, *7*, 1017-1026.
25. Filpula, D.; McGuire, J.; Whitlow, M. In *Antibody Engineering: A Practical Approach*; Chiswell, D. J.; McCafferty, J.; Hoogenboom, H., Eds.; IRL Press: Oxford, England, **1995**, submitted.
26. Essig, N. Z.; Wood, J. F.; Howard, A. J.; Raag, R.; Whitlow, M. *J. Mol. Biol.* **1993**, *234*, 897-901.
27. Desplanq, D.; King, D. J.; Lawson, A. D. G.; Mountain, A. *Protein Engng.* **1994**, *7*, 1027-1033.
28. Anand, N. N.; Mandal. S.; MacKenzie, C. R.; Sadowska, J.; Sigurskjold, B.; Young, N. M.; Bundle, D. R.; Narang, S. A. *J. Biol. Chem.* **1991**, *266*, 21874-21879.
29. Batra, J. K.; FitzGerald, D.; Gately, M.; Chaudhary, V. K.; Pastan, I. *J. Biol. Chem.* **1990**, *265*, 15198-15202.
30. Dorai, H.; McCartney, J. E.; Hudziak, R. M.; Tai, M.-S.; Laminet, A. A.; Houston, L. L.; Huston, J. S.; Oppermann, H. *Bio/Technology* **1994,** *12*, 890-897.
31. Whitlow, M.; Bell, B. A.; Feng, S.-L.; Filpula, D.; Hardman, K. D.; Hubert, S. C.; Rollence, M. L.; Wood, J. F.; Schott, M. E.; Milenic, D. E.; Yokota, T.; Schlom, J. *Protein Engng.* **1993**, *6*, 989-995.
32. Nose, M.; Takano, R.; Nakamura, S.; Arata, Y.; Kyogoku, M. *Int. Immunol.* **1990,** *2*, 1109-1112.

33. Matsuda, H.; Nakamura, S.; Ichikawa, Y.; Kozai, K.; Takano, R.; Nose, M.; Endo. S.; Nishimura, Y.; Arata, Y. *Mol. Immunol.* **1990**, *27*, 571-578.
34. Jespers, L. S.; Roberts, A.; Mahler, S. M.; Winter, G.; Hoogenboom, H. R. *Bio/Technology* **1994**, *12*, 899-903.
35. Löfdahl, S.; Guss, B.; Uhlén, M.; Philipson, L.; Lindberg, M. *Proc. Natl. Acad. Sci., USA* **1983**, *80*, 697-701.
36. Yanisch-Perron, C.; Vieira, J.; Messing, J. *Gene* **1985**, *33*, 103-119.
37. Djojonegoro, B. M.; Benedik, M. J.; Willson, R. C. *Bio/Technology* **1994**, *12*, 169-172.
38. Harlow, E.; Lane, D. *Antibodies: A Laboratory Manual*; Cold Spring Harbor Laboratory Press: Cold Spring Harbor, NY, **1988**.
39. Mason, P. J.; Williams, J. G. In *Nucleic Acid Hybridisation: A Practical Approach*; Hames, B. D.; Higgins, S. J., Eds.; IRL Press: Oxford, England, **1985**, pp. 113-137.
40. Maruyama, S.; Kubagawa, H.; Cooper, M.D. *J. Immunol.* **1985**, *135*, 192-199.
41. Lee, T. K.; Rollence, M. L.; Hallberg, P. L.; Oelkuct, M. S.; Dodd, S. W.; Nagle, J. W.; Filpula, D. R. *J. Industr. Microbiol.* **1995**, *14*, in press.

RECEIVED May 17, 1995

Chapter 7

T-Cell Targeted Immunofusion Proteins from *Escherichia coli*

Rationale, Design, and Properties

Marc Better and Patricia A. Nolan

XOMA Corporation, 1545 17th Street, Santa Monica, CA 90404

Fusion proteins between cell-targeting domains and cytotoxic proteins may be particularly effective therapeutic reagents. We constructed a family of immunofusion proteins linking the humanized Fab, F(ab')2 or single-chain antibody forms of the H65 antibody (which recognizes the CD5 antigen on the surface of human T cells) with the plant ribosome-inactivating protein gelonin. We reasoned that an immunofusion could be identified that killed human target cells at least as efficiently as a chemical conjugate of H65 and gelonin. Each immunofusion protein was produced in *E. coli* as a secreted protein and was recovered directly from the bacterial culture supernatant in an active form. All of the immunofusion proteins were purified by a common process and each immunofusion protein was tested for cytotoxicity toward antigen-positive human cells. Interestingly, a wide range of cytotoxic activity was seen among the fusion family members, and several fusion proteins were identified which are approximately as active as the most effective chemical conjugates.

Genetic fusion of genes or gene segments followed by heterologous expression has provided a wealth of information about protein function. The first examples of gene fusions were described in the 1970s and early 1980s when fusion proteins were generated to examine gene expression *in vivo* (1) and protein transport across cell membranes (2). It soon became clear that two genes could be linked to form multifunctional fusion proteins which retained at least some of the activity of both components (3, 4, 5). Currently, fusion proteins serve not only as tools to explore protein domain function, but also as tools to drive gene expression, to facilitate protein purification, to generate recombinant vaccines and to produce bifunctional therapeutic reagents. Bifunctional proteins hold particular promise as human pharmaceuticals since protein segments such as cell targeting domains can in some cases be effectively linked to active enzymes, cytotoxic proteins or specific immunogens.

In this article we describe our strategy to develop fusion proteins that can target and kill human T cells. Our recent fusion protein work evolved from earlier studies with chemically linked immunoconjugates which demonstrated that 1.) an immunoconjugate between the murine anti-human CD5 antibody H65 and the plant

0097–6156/95/0604–0086$12.00/0

toxin ricin kills human T cells *in vitro* (6) and has demonstrated clinical efficacy *in vivo* (7, 8); 2.) a humanized version of the H65 antibody (he3H65) retains high affinity for antigen (9) and its antigen binding domains Fab and F(ab')$_2$ also form effective immunoconjugates (10); and 3) the ribosome inactivating protein gelonin is an alternative to ricin since immunoconjugates with gelonin are particularly effective and resulted in a greater extent of specific cell killing than ricin immunoconjugates (11). In addition, both the he3H65 antigen binding domains and gelonin are expressed separately as recombinant proteins at high yield (approaching 1 g/L) as secreted proteins in *E. coli* (10, 12). An iterative approach was used for the development of the anti-human CD5 fusion proteins described here. We will describe the work that set the stage for our immunofusion protein program, and then describe in some detail the molecules we generated and their relevant biological properties.

Chemically linked immunoconjugates vs. fusion proteins.

Hormones, cytokines, and high affinity antibodies are ideal cell targeting agents. When properly linked to such a targeting agent, cytotoxic proteins can exert their effect at the cell surface, intracellularly post internalization or on nearby cells. Cytotoxic proteins are often coupled to cell-targeting domains via heterobifunctional crosslinking reagents, and in many cases exhibit specific cellular cytotoxicity *in vitro* and *in vivo*. Our clinical results with the H65-ricin A immunoconjugate, for example, demonstrated clinical utility for resolution of graft vs. host disease (7) and suggested promise in a variety of autoimmune indications (8, 13). Recent reviews outline the properties of targeted conjugates between cell binding domains and cytotoxic molecules (14, 15).

Conjugates between cell-targeting proteins and cytotoxic molecules are often very effective at killing target cells, but chemically linked immunoconjugates can be heterogeneous, unstable both *in vitro* and *in vivo*, and difficult to manufacture. Protein fusions between the targeting agent and the catalytic moiety can in some instances bypass these obstacles, especially if the product is produced efficiently in a microbial system. Fusion proteins currently expressed in *E. coli* are primarily designed to take advantage of the specificity conferred by their cytokine or antibody domain to target specific, and often pathogenic, receptor-bearing cell populations. The most prominent target for active fusion proteins has historically been cancer cells. The design of anti-T cell fusion proteins is also of clinical interest, and is our major area of interest.

Figure 1 illustrates the processes needed to produce a chemically-linked immunoconjugate from bacterially expressed proteins versus an immunofusion protein and shows that the production system for a fusion protein is inherently simpler than that of a chemically linked immunoconjugate. Our approach to develop the most effective products to kill CD5 antigen-positive cells was to evaluate chemical immunoconjugates and identify the most potent combinations of antigen-binding domains and cytotoxic proteins, and then to generate similar immunofusion constructs for facile production as secreted proteins in *E.coli*.

In general, the most effective chemically-linked immunoconjugates contain a labile bond between the targeting and catalytic domains. Likewise, most of the fusion proteins previously described rely on some intrinsic or introduced mechanism for release of the catalytic domains intracellularly. The mechanisms by which several well characterized fusion proteins are able to exert their specific effect on cells are illustrated below, and provide a rationale for the methodology we used to generate T cell targeted immunofusion proteins with the he3 H65 antigen binding domains and gelonin.

Fusion proteins to *Pseudomonas* exotoxin, diphtheria toxin and ricin

The best characterized and most commonly evaluated cytotoxic fusion proteins are those that direct the catalytic domains of the *Pseudomonas* exotoxin (PE) and diphtheria toxin (DT) to the surface of target cells. Most frequently, fusion proteins to these bacterial toxins are expressed in *E. coli* as insoluble proteins and refolded *in vitro* to an active form. Targeting domains have included IL-2, IL-4, IL-6, transforming growth factor alpha, insulin-like growth factor, acidic and basic fibroblast growth factors; granulocyte colony stimulating factor, CD4, melanocyte stimulating hormone alpha and a number of antibody domains which recognize subunits of the IL-2 receptor, the transferrin receptor and other tumor specific antigens (16, 17, 18, 19, 20,). We describe the mode of action for these fusion proteins to illustrate that a specific intracellular processing mechanism is involved.

The PE contains three distinct functional domains: cell targeting (domain I), intracellular translocation (domain II), and catalytic activity (domain III) (21). The available evidence suggests that the translocation and catalytic domains remain functionally active when re-directed by an alternative cell binding domain. Within the target cell, PE or fusion proteins containing domains II and III such as PE40 (amino acids 253-613 of PE) or PE38 (amino acids 253-364 and 381-613 of PE) require two essential steps to liberate the carboxyl-terminal catalytic domain capable of reaching the cytosol where it inhibits protein synthesis: (a) proteolysis between amino acids at positions 279 and 280, and (b) reduction of a disulfide bond spanning amino acids at position 265 and 287 of PE (22). A similar mechanism is also in place within the DT protein (23).

By contrast, the plant toxin ricin, which has been extensively used as a partner in chemically-linked immunoconjugates (15), lacks the functional equivalent of the PE domain II and requires introduction of an additional feature when directed to target cells as a fusion protein. Ricin is a type II RIP which in its mature form contains two polypeptide chains linked by a disulfide bond (24). The A chain contains the N-glycosidic catalytic domain, while the B chain is a lectin cell recognition domain. Available evidence with ricin A chain fusion proteins suggests that delivery to a cell alone is insufficient for catalytic activity intracellularly, as several reports have demonstrated that a mechanism for intracellular release of ricin A chain from the targeting domain must be introduced (25, 26, 27). One effective means is introduction of a segment from DT which contains the protease-sensitive and disulfide-bounded loop similar to that described above from PE. Thus, when the proper signals are all included in the fusion protein the desired biological functions can be reconstituted.

Fusion proteins to surface-acting enzymes and type-I ribosome-inactivating proteins (RIP).

The ricin fusion protein examples illustrate that it can be insufficient to simply join a cytotoxic protein and a cell targeting domain when proper intracellular processing and trafficking are required, even if both protein domains in the fusion protein retain their function *in vitro*. One strategy to circumvent this obstacle is to deliver a cytotoxic molecule to a target cell which does not require internalization for activity. This approach was used to deliver the *Clostridium perfringens* toxin to cells with an antibody Fab molecule (28). In that case, the *C. perfringens* phospholipase was fused directly to the antibody Fd chain and co-expressed with the light chain gene. The assembled Fab-fusion was effective at target cell killing .

At least some plant type I RIP, which contain only a single catalytic polypeptide, can be cytotoxic when delivered to cells as a component of a fusion protein. A recent report (29) highlights that the type I RIP saporin is active when linked directly to the basic fibroblast growth factor. Here the molecule was effective at cell killing even though the fusion protein does not retain full RIP activity *in vitro* (compared to saporin alone). Possibly with extremely active RIP such as saporin, sufficient endogenous proteolysis occurs intracellularly and the potent RIP eventually results in cellular cytotoxicity.

Gelonin Immunoconjugates kill human T cells efficiently.

Our clinical studies with an anti-human CD5 ricin A chain immunoconjugate (7, 8, 13) suggested that the CD5 target was particularly attractive for directed therapeutic reagents. We next wanted to explore new targeting vehicles including humanized antigen binding domains such as Fab and $F(ab')_2$ (to minimize any potential immune response in man), and to identify other cytotoxic proteins which might be more potent, less immunogenic and amenable to recombinant expression. We examined antibody conjugates with the a panel of type I and type II RIP to identify those cytotoxic molecules that formed conjugates with the highest specific cytotoxicity and the lowest non-specific toxicity (S. Bernhard and S. Carroll, XOMA Corporation, in preparation). Of the molecules tested, the most promising immunoconjugates were with the type I RIP gelonin (~30 kDa) which is known to be relatively non-toxic.

The gelonin cDNA gene was then cloned from *Gelonium multiflorum* seeds, and a system was developed for expression of recombinant gelonin (rGel) as a secreted protein in *E. coli* (12). Interestingly, rGel was expressed at high yield (> 1 g/L) unlike some other type I RIP, and could be recovered from the culture supernatant easily (12). The rGel was as active as native gelonin (nGel) in an *in vitro* assay for protein synthesis inhibition, Table I.

Table I. Inhibition of protein synthesis *in vitro*

RIP	IC_{50} (pM)
RTA_{30}	3
nGel	15
rGel	11
$rGel_{K10C}$	60
$rGel_{C50A}$	20
$rGel_{C44A}$	47
$rGel_{N60C}$	26
$rGel_{N239C}$	955
$rGel_{K244C}$	32
$rGel_{D247C}$	12
$rGel_{K248C}$	47
$rGel_{C44A/C50A}$	16
$rGel_{K10C/C44A/C50A}$	7
$rGel_{D247C/C44A/C50A}$	20

Protein synthesis inhibition was measured as described (10) By comparison with uninhibited samples, the concentration of toxin (pM) which inhibits protein synthesis by 50% (IC_{50}) was calculated. IC_{50} values of 15 and 11 are identical within the variability of this assay.

Immunoconjugates constructed *in vitro* with rGel were tested for cytotoxicity to human T cell lines and human peripheral blood mononuclear cells, Table II. The HSB2 cytotoxicity assay measures relative immunofusion cytotoxicity with a laboratory cell line, while the PBMC assay is a paradigm for immunofusion cytotoxicity on human T cells likely to be similar to those involved in human disease.

Table II. Activity of H65-gelonin immunoconjugates on HSB2 cells and PBMC

Conjugate	**HSB2 IC50 (pM RIP)**	**PBMC IC50 (pM RIP)**
Fab'-rGel	135	160
F(ab')$_2$-rGel	55	34
IgG-rGel	270	50
IgG-nGel	1770	150
Fab'-rGel$_{D247C}$	48	210
F(ab')$_2$-rGel$_{D247C}$	23	32
IgG-rGel$_{D247C}$	41	49
Fab'-RTA$_{30}$	530	2200
F(ab')$_2$-RTA$_{30}$	16	90
IgG-RTA$_{30}$	40	270

Antibody conjugates to rGel, nGel and rGel$_{D247C}$ were prepared as described (10), and antibody conjugates to RTA$_{30}$ were prepared as described (35)

Typically, immunoconjugates between antibodies and the A chain of type II RIP such as ricin are formed between antibodies into which a reactive thiol is introduced with a heterobifunctional cross linking agent and the native cysteine thiol of the A chain. Gelonin, by contrast, contains two cysteine residues which are involved in an intramolecular disulfide bond, but does not contain an available thiol for conjugation to antibody (10). Gelonin is easily inactivated by chemical modification with crosslinking agents and we wanted to genetically engineer gelonin to minimize potentially destructive modification and simplify the conjugation procedure. We therefore engineered gelonin to generate a family of gelonin analogs, each with a single unpaired cysteine residue (10). The amino acid positions chosen for modification to cysteine coincide with surface-accessible loops in the probable three dimensional structure of gelonin (based on the 35% amino acid identity with ricin A chain (12), the 36% amino acid identity to trichosanthin (12), and the almost superimposable backbone structure of ricin and trichosanthin (30, 31)), or with the positions of endogenous cysteine residues. In most cases, enzymatic activity *in vitro* was unaltered by this modification, Table I . The rGel analogs were conjugated to H65 IgG, Fab and F(ab')$_2$ via their unpaired cysteine residue. Several rGel analogs formed immunoconjugates that were more cytotoxic to antigen-bearing cells than those made with linker-modified rGel, Table II, whereas others were less potent (10).

Design of gelonin immunofusion proteins.

Since both components (gelonin and antigen binding domains) were produced as secreted proteins and both appeared as fully active, folded proteins in the *E. coli* culture supernatant, we expressed fusion proteins between antigen-binding domains and gelonin as secreted proteins. The results with gelonin chemical conjugates described in the previous section illustrated the importance of the proper spatial arrangement between gelonin and its targeting agent for optimal activity. We therefore constructed a family of similar fusion proteins to identify the optimal domain orientation. The humanized variable region genes for the H65 antibody, he3 (9), and the gelonin gene (12) served as the starting materials for the construction of these T cell-targeted immunofusion genes. We generated gene fusions that would contain antibody targeting domains in three formats: Fab, $F(ab')_2$, and single chain antibody (SCA). It was initially unclear how the binding domain avidity, or the specific linkage of independently folding components would affect the final activity. Each family member was expressed in *E. coli* using a previously described expression system (32), and purified from the culture broth.

Two SCA versions of the he3H65 antibody (V_LV_H and V_HV_L) were constructed by overlap extension PCR (33) from the he3H65 V_k and V_H genes. Oligonucleotide primers introduced the $[(Gly)_4Ser]_3$ inter-domain linker (34) and the gelonin gene was fused in-frame to each SCA at either the 5' or 3' end. The gelonin gene was likewise linked to the 5'-end of the V_L-Ck or V_H-C_H1 encoding sequence, and the fusion genes were assembled into a dicistronic message with the cognate Fd- or k-encoding sequence respectively. The gelonin-kappa fusion gene was also incorporated into a dicistronic message with an Fd' gene which encoded both IgG1 interchain hinge cysteine residues and the first 9 amino acids of C_H2. Inclusion of this segment allows direct *E. coli* expression of the divalent $F(ab')_2$, (35), and $F(ab')_2$ -fusion proteins were expressed.

We engineered a possible intracellular release mechanism into fusion proteins by introducing one of two short peptide sequences between the antigen targeting domain of the humanized H65 antibody and gelonin, Figure 2. These peptide segments from the *E. coli* shiga-like toxin (SLT, 36) and rabbit muscle aldolase (RMA, 37) are 20 amino acids in length. The SLT sequence contains a disulfide bounded peptide with a recognition site for trypsin-like proteases and resembles the cleavable disulfide loop of PE and DT, while the RMA sequence contains several sites that are susceptible to the lysosomal enzymes Cathepsin B and Cathepsin D (37). We reasoned that these peptides were likely to be cleaved intracellularly resulting in gelonin release. Direct fusions without either the SLT or RMA linker peptides were also constructed. Figure 3 schematically illustrates the immunofusion proteins we produced.

Production of fusion proteins.

A plasmid vector containing each fusion gene was transformed into *E coli*, and the bacterial cultures were grown in a 10 liter fermenter. When each culture reached an optical density (OD_{600}) of about 100, the cells were induced with L-arabinose. Each culture was harvested 20 to 24 hours post induction, and the cells were separated from the supernatant (which contains the recombinant protein).

The resultant fusion proteins were purified directly from the *E. coli* fermentation broth by a series of chromatographic steps. A single purification method was developed for all immunofusions, Figure 4. This procedure was sufficient to recover milligrams of fusion protein from each fermentation batch. The immunofusion proteins were generally greater than 80% pure. The production and purification of a representative immunofusion example is shown in Figure 5.

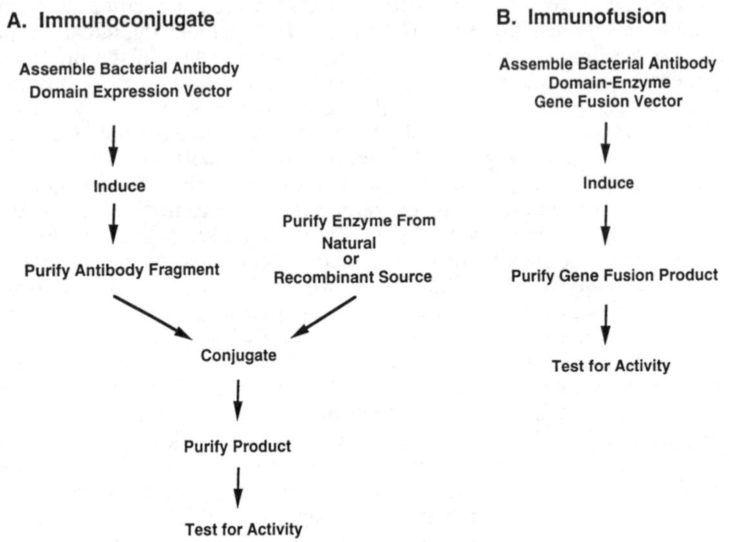

Figure 1. Process for production of chemically-linked immunoconjugates (A), and immunofusion proteins (B).

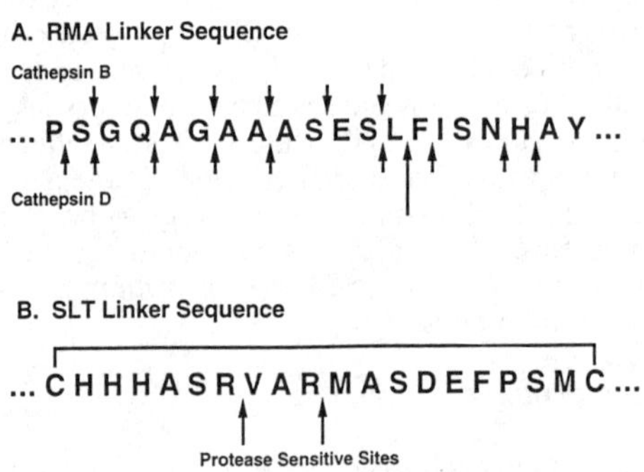

Figure 2. Amino acid sequence of peptide linkers from rabbit muscle aldolase (RMA, 37) and the *E. coli* shiga-like toxin (SLT, 36), panels A and B respectively. Amino acid site likely to be susceptible to proteolysis are indicated with arrows.

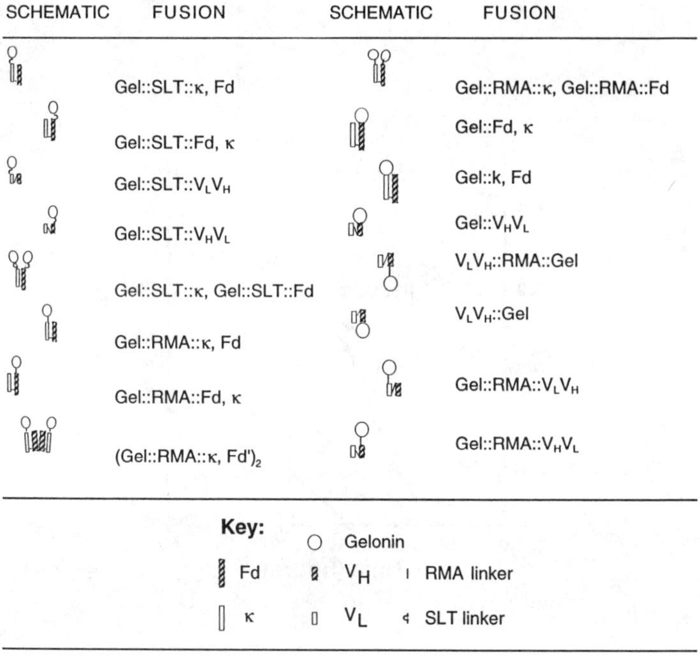

Figure 3. Gelonin fusion proteins produced from *E. coli.*

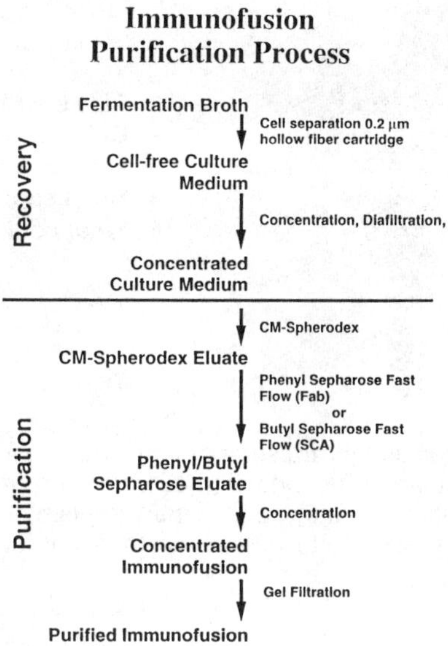

Figure 4. Steps used to purify immunofusion proteins from bacterial fermentation broth.

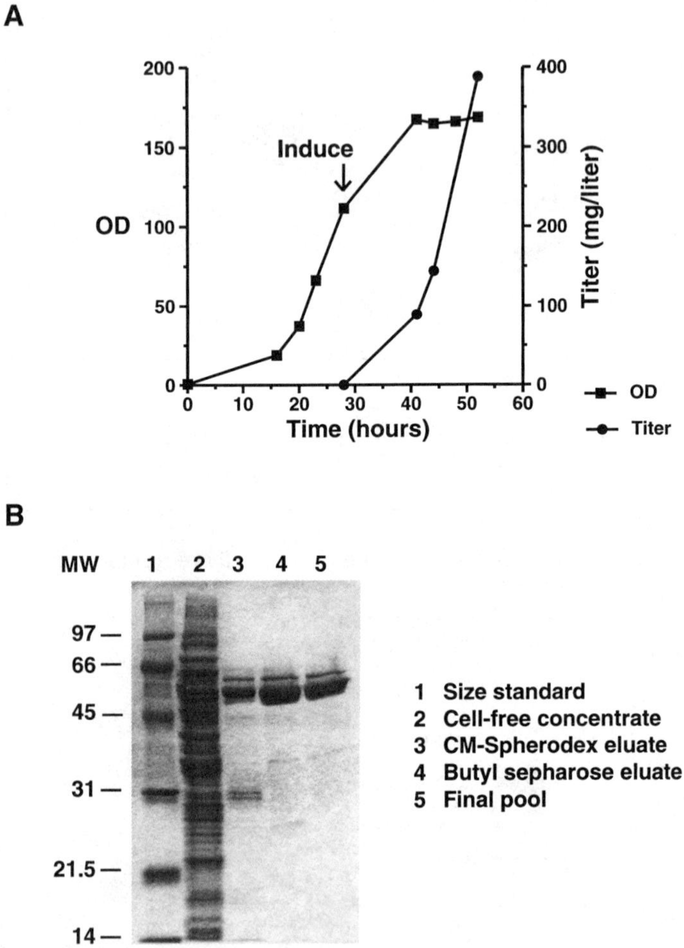

Figure 5. Expression and purification of a SCA-gelonin fusion protein. A. Fusion protein accumulates after induction of fermenter-grown *E. coli* with arabinose. B. The Coomassie-stained SDS-polyacrylamide gel illustrates the purification of fusion protein following the protocol outlined in Figure 5.

Affinity of Fusion Proteins.

Several of the fusion proteins were compared to intact IgG and Fab in a competition binding assay for the CD5 antigen on the surface of MOLT-4 cells, Table III. The F(ab')$_2$-fusion protein retains roughly half of the binding affinity of the IgG while the Fab-fusion proteins retain roughly half the affinity of the Fab. The SCA fusion proteins tested had a binding affinity about 3 to 10 times lower than the Fab but had an affinity roughly equivalent to the he3H65 SCA.

Table III. Affinity of Fusion Proteins for the CD5 antigen on Molt-4 cells.

Antigen-binding protein	Kd (nM) 4°C	Kd (nM) 37°C
he3 IgG	0.8	1.1
he3 Fab	2.6	5.4
Gel::RMA::k, Gel::RMA::Fd	11.8	23
Gel::RMA::k, Fd' (monomer)	5.8	10.8
Gel::RMA::k, Fd' (dimer)	1.9	1.2
SCA::RMA::Gel	9.5	17

Comparison of IgG and Fab to immunofusions. Assays and analysis are as described (9, 35).

Activity of fusion proteins.

Each purified fusion protein was tested for specific cytotoxicity against the human T cell line HSB2 and purified human peripheral blood mononuclear cells (PBMC). Figure 6 highlights the activity of four fusion proteins with the RMA linker on HSB2 cells. As expected, some fusion proteins are more cytotoxic than others. Fusions containing the SCA at either the N- or C-terminus of the molecule appear to be equally effective at killing cells, and there does not seem to be a clear advantage to fusion proteins containing Fab or SCA. A striking difference is seen however, between monovalent and divalent forms of the fusion proteins. The (Gel::RMA::k, Fd')$_2$ molecule is roughly 10 times more effective at cell killing than the monovalent form. In general, fusions containing the SLT linker are more cytotoxic to HSB2 cells than are fusions containing the RMA linker or no linker at all.

Several clear patterns of cytotoxicity emerged from the assays with PBMC. In a comparison among RMA linked fusion to SCA, Fab, and F(ab')$_2$, the divalent immunofusion is again clearly the most potent, Table IV. The Fab and SCA fusions are approximately of equal activity. The peptide linker between functional domains of the fusion proteins appears relatively unimportant for activity against PBMC, however, since the fusions which lack either the SLT or RMA linkers retain activity (data not shown). The immunofusion without a specific linker may contain an amino acid sequence at the interdomain junction that creates a susceptible cleavage site or, alternatively, these gelonin immunofusion proteins may be transported to the cytoplasm of the cells intact and remain in an active form. The single chain fusions with gelonin at the C-terminus seem to be more effective at cell killing than those with gelonin at the N-terminus, suggesting that linkage too close to the antigen binding site is disadvantageous.

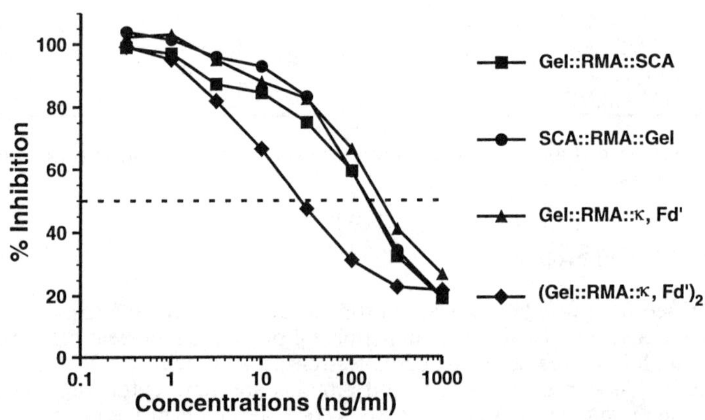

Figure 6. Cytotoxicity of four immunofusion proteins on HSB2 cells. The dashed line represents the point at which protein synthesis is inhibited by 50%.

Table IV. Cytotoxic Potency on PBMC

	IC_{50} (ng/ml) Median	% Inhibiiton
CD5 Plus (H65-RTA)	98	60 ± 11
he3 Fab-Gel_{C50A}	13	92 ± 5
(Gel::RMA::k, Fd')$_2$	5.7	95 ± 1.3
(Gel::RMA::Fd', k)$_2$	6.6	95 ± 2.1
Gel::RMA::Fd, Gel::RMA::k	28	92 ± 12
Gel::RMA::Fd, k	69	88 ± 4.2
SCA::RMA::Gel	67	95 ± 2.9
Gel::RMA::SCA	130	89 ± 6.6

Comparison of several RMA immunofusions to chemical conjugates. Assays were performed as described (10, 11, 35). % Inhibition is the percent inhibition of macromolecular synthesis at the highest reagent concentration tested.

Conclusions

Multifunctional fusion proteins combining enzymes and targeting domains may have many pharmaceutical and diagnostic applications. Two important considerations for drug development are how efficiently these molecules fold after expression (yield) and how much of the individual domain function is retained (activity). We reasoned that introduction of a specific intracellular cleavage mechanism such as that found in PE and DT fusion proteins would be necessary for maximal cytotoxicity of gelonin fusion proteins. This seemed likely since ricin immunofusion proteins were essentially inactive in the absence of such a mechanism (25, 26, 27). Our data suggest, however, that gelonin fusions with an engineered protease-sensitive linker are not much more cytotoxic to human PBMC than those without such a linker. An intriguing possibility is that unlike the type II RIP ricin, gelonin (and possibly other type I RIP) has evolved a distinct mechanism for intracellular cytotoxicity that does not require specific separation from any delivery agent. Evidence to suggest that type I and type II RIP behave differently intracellularly when fused to a targeting agent has been presented since FGF-saporin fusion proteins are cytotoxic to receptor-positive cells (29). Additional experiments will be required to determine this point.

We are interested in reagents that can specifically kill the CD5-positive cells implicated in human disease. Previous work with chemically linked immunoconjugates had demonstrated that no particular RIP is likely to form the most effective conjugate with all cell-targeting molecules, and we found that the most cytotoxic conjugates with the anti-human CD5 antibody H65 are those with gelonin The gelonin immunofusion molecules described here exhibit a wide range of cytotoxic activity. This range may not be unexpected, however, since individual members of the immunofusion family differ in antigen affinity, and are likely to differ in protein stability and the ability to be internalized and processed into an active gelonin catalytic domain intracellularly. The most important conclusions from the work described here are that H65-gelonin immunofusion proteins can be as cytotoxic to human PBMC as the H65-ricin A chain immunoconjugate which has seen wide clinical use, and as cytotoxic as the most effective chemical immunoconjugates between H65 antigen binding domains and recombinant gelonin, Table IV. Importantly, these fusion proteins can be prepared at high yield directly from the supernatant of induced *E. coli* cultures. Additional studies are required to evaluate the clinical potential of these fusion protein products.

Acknowledgments

We thank Drs. Susan L. Bernhard, Stephen F. Carroll, Dianne M. Fishwild, Shau-Ping Lei and Rob Robinson, and Bob Williams, Darryl Garrison, Wilfredo Morales, Patrick Gavit, Manik Baltaian, Maria Molina, Sandra Soares, Anne Orme, and Hsui-Mei Wu for making this work possible.

Literature Cited

1. Casadaban, M.J., Chou, J., and Cohen, S.N. *J. Bacteriol.* **1980**, *143*, 971-980.
2. Silhavy, T.J., Shuman, H.A., Beckwith, J., and Schwartz, M. *Proc. Natl. Acad. Sci. U.S.A.* **1977**, *74*, 5411-5415.
3. Uhlén, M., Nilsson, B., Guss, B., Lindberg, M., Gatenbeck, S., Lennart, P. *Gene (Amsterdam)* **1983**, *23*, 369-378.
4. Mata-Gilsinger, M. and Ritzenthaler, P. *Gene (Amsterdam)* **1983**, *25*, 9-20.
5. Sharon, J., Gefter, M.L., Manser, T., Morrison, S.L., Oi, V.T., and Ptashne, M. *Nature* , **1984**, *309*, 365-367.
6. Kernan, N.A., Knowles, R.W., Burns, M.J., Broxmeyer, H.E., Lu, L., Lee, H.M., Kawahata, R.T., Scannon, P.J., and Dupont, B. *J. Immunol.* **1984**, *133*, 137-146.
7. Byers, V., Henslee, C.F., Kernan, N., Blazar, B.R., Gingrich, R., Phillips, G.L., LeMaistre, C.F., Gilliland, G., Antin, J.H., Vogelsang, G., Martin., P, Tutscha, P.J., Trown, P., Ackerman, S.K., O'Reilly, R.J., Scannon, P.J. *Blood* **1990** *75*, 1426-1432.
8. Strand, V., Lipsky, P.E., Cannon, G.W., Calabrese, L.H., Wiesenhutter, C., Cohen, S.B., Olsen, N.J., Lee, M.L., Lorenz, T.J., and Nelson, B *Arthritis and Rheumatism* **1993**, *36*, 620-630.
9. Studnicka, G., Soares, S., Better, M., Williams, R.E., Nadell, R., and Horwitz, A.H. *Protein Engineering* **1994**, *7*, 805-814.
10. Better, M., Bernhard, S.L., Fishwild, D.F., Nolan, P.A., Bauer, R.J., Kung, A.H.C., and Carroll, S.F. *J. Biol. Chem.* **1994**, *269*, 9644-9650.
11. Fishwild, D.M., Wu, H.-M., Carroll., S.F., and Bernhard, S.L. Clinical & *Experimental Immunol.* **1994**, in press.
12. Nolan, P.A., Garrison, D.A., and Better, M. *Gene (Amsterdam)* **1993**, *134*, 223-227.
13. Skyler, J.S., Lorenz, T.J., Schwartz, S., Eisenbarth, G.S., Einhorn, D., Palmer, J.P., Marks, J.B., Greenbaum, C., Saria, E.A., and Byers, V. J. *Diabetes. Comp.* **1993**, *7*, 224-232.
14. Pastan, I. and FitzGerald, D. *Science* **1991**, *254*, 1173-1177.

15. Vitetta, E.S., Thorpe, P.E., and Uhr, J.W. *Immunol. Today* **1993**, *14*, 252-259.
16. Pastan, I., Chaudhary, V. and FitzGerald, D.J. *Annu. Rev. Biochem.* **1992**, *61*, 331-354.
17. Williams, D.P., Parker, K., Bacha, P., Bishai, W., Borowski, M., Genbauffe, F., Strom, T.B. and Murphy, J.B. *Protein Engineering* **1987**, *1*, 493-498.
18. Gawlak, S.L., Pastan, I. and Siegall, C.B. *Bioconjug. Chem.* **1993**, *4*, 483-489.
19 Chadwick, D.E., Williams, D.P., Niho, Y., Murphy, J.R. and Minden, M.D. *Leuk. Lymphoma* **1993**, *11*, 249-262.
20. Chaudhary, V.K., Queen, C., Junghans, R.P., Waldman, T.A., FitzGerald, D.J. and Pastan, I. *Nature* **1989**, *339*, 394-397.
21. Hwang, J., FitzGerald, D.J., Adhya, S., and Pastan, I. *Cell* **1987**, *48*, 129-136.
22. Ogata, M., Fryling, C.M., Pastan, I., and FitzGerald, D.J. *J. Biol. Chem.* **1992**, *267*, 25396-25401.
23. Iglewski, B.H., and Kabat, D. *Proc. Natl. Acad. Sci. U.S.A.* **1975**, *72*, 2284-2288.
24. Montford, W., Villafranca, J.E., Monzingo, A.F., Ernst, S.R., Katzin, B., Ruterber, E., Xuong, N.H., Hamlin, R., and Robertus, J.D. *J. Biol. Chem.* **1987**, *262*, 5398-5403.
25. O'Hare, M., Brown, A.N., Hussain, K., Gebhardt., A., Watson, G., Roberts, L.M., Vitetta, E.S., Thorpe, P.E., and Lord, J.M. *FEBS Lett.* **1990,** *273*, 200-204.
26. Westby, M., Argent, R.H., Pitcher, C., Lord, J.M., and Roberts, L.M. *Bioconjugate Chem.* **1992**, *3*, 375-381.
27. Cook, J.P., Savage, P.M., Lord, J.M., and Roberts, L.M. *Bioconjugate Chem.* **1993**, *4*, 440-447.
28. Chovnick, A., Schneider, W.P., Tso, J.Y., Queen, C., and Chang, C.N. *Cancer Research* **1991**, *51*, 465-467.
29. Lappi, D.A., Ying, W.B., Barthelemy, I., Martineau, D., Prieto, I., Benatti, L., Soria, M., and Baird, A. *J. Biol. Chem.* **1994**, *269*, 12552-12558.
30. Katzin, B.J., Collins, E.J., and Robertus, *Proteins* **1991**, *10*, 251-259.
31. Zhou, K., Fu, Z., Chen, M., Lin, Y., and Pan, K. *Proteins* **1994,** *19*, 4-13.
32. Better, M. and A. Horwitz. In *Antibodies, antigens, and molecular mimicry*, J.J. Langone, Ed.; *Methods in Enzymology* Academic Press: San Diego, CA 1989, Vol 178; pp 476-496.
33. Ho, S.N., Hunt, H.D., Horton, R.M. Pullen, J.K., and Pease, L.R. *Gene (Amsterdam)* **1989**, *77*, 51-59.
34. Huston, J.S., Levinson, D., Mudgett-Hunter, M., Tai, M-S., Novotny, J., Margolies, M.N., Ridge, R.J., Bruccoleri, R.E., Haber, E., Crea, R., and Oppermann, H. *Proc. Natl. Acad. Sci. U.S.A.* **1988**, *85*, 5879-5883.
35. Better, M., Bernhard, S.L., Lei, S-P., Fishwild, D.M., Lane, J., Carroll, S., and Horwitz, A.H. *Proc. Natl. Acad. Sci. U.S.A.* **1993**, *90*, 457-461.
36. Calderwook, S.B., Auclair, F., Donohue-Rolfe, A., Keusch, G.T., and Mekalanos, J.J. *Proc. Natl. Acad. Sci. USA* **1987***84*, 4364-4368.
37. Kirschke, H., and Barrett. in '*Lysosomes: Their role in protein breakdown.* Glaumann, H. and Ballard, F.J., Eds. Academic Press, Orlando, FL; pp. 193-238.

RECEIVED May 17, 1995

Chapter 8

Production and Purification of Antibody Using *Bacillus subtilis* as an Expression Host

S. L. Wong[1], X. C. Wu[1], L. P. Yang[1], S. C. Ng[2], and N. Hudson[3]

[1]Department of Biological Sciences, University of Calgary, Calgary, Alberta T2N 1N4, Canada
[2]Abbott Laboratories, Abbott Park, IL 60064–3500
[3]Northwest Center for Medical Education, Indiana University Medical Center, Gary, IN 46408–1197

Bacillus subtilis serves as an attractive expression host for the production and secretion of single-chain antibody derivatives. With the recent developments in constructing extracellular protease deficient strains and secretion vectors, production of stable and biologically active single-chain antibody fragments is technically feasible. Introduction of a histidine tag at the C-terminus of an antidigoxin single-chain antibody fragment simplifies the purification of this protein without reducing the production yield. Although the antidigoxin single-chain antibody precursor tends to form inclusion bodies, coexpression of a set of molecular chaperones (GroES/GroEL) via a binary expression vector system allows a 2.5-fold increase in the production yield of this protein. It is possible to further improve the production yield through genetic manipulations and fermentation studies.

With the recent advances in genetic and protein engineering, various novel antibody derivatives including single-chain antibody (SCA) fragments (*1-2*) humanized antibodies (*3-4*) immunotoxins(*5*) antibody-metallothionein fusions (*6-7*) bispecific antibodies (*8*) miniantibodies (*9*) and catalytic antibodies (*10-11*) can be designed. With these creative ideas in designing antibodies, it is vital that there should be expression systems available to produce the designed molecules in a biologically active form. With the pioneering works from Boss (*12*), Pluckthun (*13*), Better (*14*), and Winter (*3*), E. *coli* is widely accepted as an expression host for producing various forms of engineered antibody fragments. Since E. *coli* is used in most of the recombinant DNA works and there are many vectors and strains available, all these factors contribute to make E. *coli* an attractive system to work with. In most of the cases, antibody derivatives are produced in E. *coli* intracellularly and the formation of inclusion bodies is fairly common. To produce soluble and functional antibody fragments without tedious and unpredictable refolding of these inclusion bodies, export of these proteins to the periplasmic space in E. *coli* is a method of choice. Protein export is particularly important for the functional production of Fv since the variable regions of light chain and heavy chain can be effectively concentrated in a small volume and assembled efficiently into Fv. Furthermore, the nonreducing environment in the perplasmic space allows

0097–6156/95/0604–0100$12.00/0

disulfide bond formation which is vital to generate functional Fv and its derivatives (*15-16*). However, certain antibody fragments were unable to translocate through the E. *coli* inner membrane (*17*) or the translocated protein showed certain unexpected properties (*18*). Therefore, it would be ideal to have other procaryotic expression systems available to produce these engineered antibody fragments in their biologically active forms. *Bacillus subtilis*, a gram-positive microorganism, is an attractive alternative.

Advantages of Using *B. subtilis* as an Expression Host

B. subtilis has many attractive features to serve as an expression host (*19-22*). Unlike E. *coli*, *B. subtilis* has only one cytoplasmic membrane. Any secretory proteins that can translocate through this membrane will be released directly to the culture medium. This feature can greatly simplify the purification process since *B. subtilis* cells can be removed from the culture by simple centrifugation. The protein of interest can be purified from other proteins in the culture medium by a few chromatographic steps without any complications from the *B. subtilis* intracellular proteins. Since B. subitlis is known to secrete many extracellular enzymes in large quantities (*23*) production of single-chain antibody fragments via secretion would be very attractive. *B. subtilis* is nonpathogenic and therefore would be suitable for producing heterologous proteins for medical applications. As the best biochemically and genetically characterized gram-positive microorganism, requirements for optimum gene expression and efficient plasmid transformation are well established (*24*). Furthermore, without a significant bias in codon usage (*25*) translation of foreign proteins can proceed smoothly in *B. subtilis* once translation initiation begins. No particular optimization of the coding sequence would be required. Lastly, large-scale fermentation methods for *Bacillus subitlis* are well established (*26*). It is possible to use various fermentation techniques to improve the final production yield.

Strain Improvement

One of the major barriers preventing the successful us of *B. subtilis* as an efficient host for the production of foreign proteins is the secretion of a high level of extracellular proteases from *B. subtilis*. Currently, *B. subtilis* is known to produce at least seven extracellular proteases. The two major proteases, natural protease A and alkaline protease (also known as subtilisin) account for 90% of the total extracellular protease activity. These extracellular proteases have a wide spectrum of substrate specificity and can rapidly degrade many secreted foreign proteins. With extensive efforts in cloning these protease genes (*27-28*) all seven structural genes encoding extracellular protease have been cloned and characterized. With the inactivation of chromosomal protease genes through both deletion and insertion, a series of extracellular protease deficient strains has been constructed (*29*). WB600, a six protease deficient *B. subtilis* strain, has been well characterized (*30*). It has only 0.3% of the wide-type extracellular protease activity. By using the TEM-β-lactamase as a marker, the half-life of this protein in B.subtilis extends significantly form 1.5 hours with the wild type *B. subtilis* 168 as the host to 40 hours with WB600 as the production host. With the addition of 2 mM phenylmethanesulphonyl fluoride (a serine protease inhibitor) to the WB600 culture, the half-life of the TEM-β-lactamase can extend further to 80 hours. A seven extracellular protease deficient strain, WB700, has recently been constructed and it has only 0.1% of the wild type extracellular protease activity (Yang, L.P. and Wong, S.L., The University of Calgary, unpublished data). All these protease-deficient strains grow normally in both the rich and defined media. The use of WB600 has recently been shown to enhance both the stability and the production yield of foreign proteins including an antidigoxin single-chain antibody fragment (*31*) and streptokinase (*32*). Construction of these protease deficient strains makes the application of *B. subtilis* as an efficient expression host for foreign secretory proteins possible.

Design of Secretion-Expression Vectors

Two expression vectors are designed to simplify the expression of foreign genes. All these vectors are derivatives of pUB110, a high copy number plasmid (about 80 copies per cell). For the first vector pWB705, Figure 1, a strong, constitutively expressed *B. subtilis* promoter, P43 (*33*) is installed in the vector to drive the transcription of foreign genes while an idealized signal peptide sequence (*34*) derived from the *B. subtilis* levansucrase gene is also installed downstream from the P43 promoter to direct secretion. Multiple cloning sites are introduced downstream of the signal peptidase cleavage to facilitate the insertion of foreign genes and to adjust the reading frame. To produce secretory proteins with authentic N-terminal sequences, a unique HindIII site is introduced upstream of the sequence encoding the signal peptidase cleavage site. Through PCR, nucleotide sequence carrying the HindIII site and the signal peptidase cleavage site can be added upstream to the sequence encoding the authentic N-terminal sequence of the foreign protein. After insertion of the foreign gene at the HindIII site, clones with the HindIII fragment inserted in the correct orientation can be selected by proper restriction digestions. Production of mature foreign proteins with authentic N-terminal sequence is expected after the proper removal of the levansucrase signal sequence by the *B. subtilis* signal peptidase. In our studies, production of an antidigoxin single-chain fragment (*35*) and streptokinase (*32*) with their authentic N-terminal sequences can indeed be achieved. The second vector is the same as pWB705 except that the constitutively expressed P43 promoter is replaced by a sucrose inducible promoter from the *B. subtilis* levansucrase (*sacB*) gene (*36*). If constitutive expression of foreign genes causes toxicity problems to the cell, the use of the sucrose inducible vectors can potentially overcome this problem.

Production of an Antidigoxin Single-Chain Antibody Fragment from *B. subtilis*

To test the *B. subtilis* expression system for its ability to produce a functional single-chain antibody fragment, plasmid pATD2 was constructed Figure 2. This plasmid carries a sequence encoding the antidigoxin single-chain antibody fragment. The sequence of this single-chain antibody fragment is essentially derived from the cDNA clones of the mouse monoclonal antibody 26-10 (*37*). Its expression is controlled by the P43 promoter and secretion is mediated with the aid of the levansucrase signal sequence. *B. subtilis* WB600 carrying pATD2 produces an extra protein band with the apparent molecular mass of 29-kDa as the culture supernatant from these cells was analyzed by SDS-polyacrylamide gel electrophoresis. This protein is confirmed to be the antidigoxin-SCA by N-terminal sequence analysis. Sequence of the first five amino acids from this protein matches exactly with that from the light chain of the 26-10 monoclonal antibody. This demonstrates the proper removal of the levansucrase signal sequence by the *B. subtilis* signal peptidase. Furthermore, this 29-kDa protein can selectively bind to the ouabain-sepharose matrix and be eluted off from the affinity column by ouabain. Affinity and specificity (with a set of digoxin analogs) determinations indicate that this antidigoxin SCA retains similar affinity and specificity as the intact 26-10 monoclonal antibody (*35*). All the secreted antidigoxin SCA are biologically active since they all can bind to the ouabain-Sepharose column and can be eluted from the column by ouabain. It is important to use the 6-protease deficient strain WB600 as the host since the stability of the secreted antidigoxin SCA increases significantly. The half-life of the *B. subtilis* WB600 produced SCA in the bacterial culture is around 32 hours. Its stability is comparable to that of the E. *coli* TEM β-lactamase produced from WB600. However, the half-life of the secreted antidigoxin SCA produced from the wild type strain is only 2.5 hours. In fact, production of a SCA fragment in *B. subtilis* was briefly reported in 1989 (*38*) before the development of the 6-protease deficient strain WB600. The stability and production yield of this SCA protein was not mentioned.

Affinity Purification of the Secreted SCA Proteins

To take full advantage of the capability of *B. subtilis* in secreting SCA fragments directly to the culture medium, it would be ideal to purify the secreted SCA fragments by an affinity chromatography. Since metal chelate chromatography is a simple and powerful tool for affinity purification of proteins with a histidine tag or other metal binding sites, it would be attractive to apply this strategy to purify the secreted SCA fragments (*39-40*). To achieve this objective, a short sequence encoding five histidine residues was added to the 3' end of the antidigoxin SCA gene via PCR. Western blot analysis indicates that the addition of five histidine residues at the C-terminal region of the antidigoxin SCA does not affect the production yield. To minimize non-specific adsorption, protein samples for metal chelate chromatography are in general stored in a high salt buffer. Therefore, the secreted SCA-$(his)_5$ proteins from the culture supernatant have to be concentrated first by ammonium sulfate precipitation. After extensive dialysis to remove ammonium sulfate and to equilibrate the protein sample with the loading buffer, the protein sample can be applied to the Ni (II)-immobilized affinity column. Antidigoxin SCA-$(his)_5$ can be separated from other proteins by eluting the bound material with an imidazole gradient (0-100 mM). Addition of five histidine residues at the C-terminus does not affect both the binding affinity and specificity of the antidigoxin SCA to digoxin and its analogues. This method can potentially be applied to affinity purified other secreted SCA-$(his)_5$ molecules from *B. subtilis*.

Inclusion Body Formation in *B. subtilis*

The production yield of the secreted antidigoxin SCA fragment in this *B. subtilis* system cultured in a regular shake flask culture is about 5 mg per liter. This production level is comparable to or even higher than some of the SCA fragments produced from E. *coli* (via the secretory pathway) in a shake flask culture. To enhance the production yield, it would be important to identify the rate-limiting step(s) in this production process. Analyzing the total protein profile from the cell pellet of WB600[pATD2] demonstrated that a high level of antidigoxin SCA fragments could be found intracellularly. To understand why these SCA fragments cannot be secreted, total intracellular proteins were further fractionated into soluble and insoluble fractions and almost all the intracellular SCA proteins were found in the insoluble fraction. Therefore, formation of inclusion bodies may likely reduce the production yield of the secreted antidigoxin SCA. This observation has two implications: (1) the antidigoxin SCA precursor has a potential to form includion bodies; (2) certain steps (such as secretion) after the biosynthesis of SCA precursors may be rate-limiting. Intracellular accumulation of these proteins may promote the formation of inclusion bodies. To address these problems, sequence of the SCA precursors can be modified through site-directed mutagenesis if critical residues in promoting aggregation without affecting biological activities can be identified. The resulting muteins may have a lower tendency to form inclusion bodies. Co-expression of molecular chaperones at a higher level may also help maintaining SCA precursors in a soluble or translocation competent form. The latter approach is attractive since it equips cells to have a better capability to cope with problems associated with protein folding and the formation of inclusion bodies. Cytoplasmic chaperone *SecB* in E. *coli* is shown to maintain some of the exported proteins in a translocation competent conformation. However, no *SecB* homolog has been found in *B. subtilis* at the present time. On the other hand, cytoplasmic molecular chaperone complexes formed by GroES and GroEL have been shown in E. *coli* to promote the export of LamB-LacZ fusion proteins (*41*) and are required for efficient secretion of β-lactamase (*42*). Overproduction of GroEs/GroEL can also enhance the solubility of procollagenase precursors produced in E. *coli* (*43*). With the independent works from Schumann (*44*) and our research group (*45*) in cloning and characterizing the *B. subtilis* groE operon, it would be interesting to see whether overproduction of GroES and GroEL can increase the final production yield of the antidigoxin-SCA fragment.

Development of a Binary Vector System for SCA Production

To achieve the co-expression of groE with the antidigoxin SCA gene, groE can be inserted either into pATD2 or into another plasmid that can co-exist with pATD2. Development of a binary vector system would be a better choice since it greatly simplifies the plasmid construction and offers expandability in the future for installing other structural genes encoding molecular chaperones, components involved in the secretory pathway and transcriptional activators. With this idea, one vector carries the gene of interest for expression studies. In this case, the vector is pATD2. The other vector should produce accessory factors required for enhanced production of SCA proteins. As shown in Figure 3, a pE194 (cop6) derivative (*46*) is constructed for installing a sucrose inducible groE operon (*45*). This plasmid is compatible with the pUB110 replicon in pATD2 and can be selected by using erythromycin and lincomycin in the culture medium.

Enhanced Production of the Antidigoxin SCA Fragment with the Aid of GroES/GroEL

By using the binary vector system, secretion of the antidigoxin SCA was found to enhance 2.5-fold. With the synthesis of SCA inside the cell, SCA fragments can be directed to three possible routes. They can translocate through the membrane. If these proteins accumulate inside the cell, they can either be degraded by intracellular proteases or can form inclusion bodies. If GroES/GroEL can enhance the solubility of the SCA precursors as observed in the case of procollagenase (*43*) minimize the degradation of SCA precursors or maintain SCA precursors in a translocational competent conformation, all these actions can lead to a higher level of SCA production. Identification of the actual mechanism(s) of GroES/GroEL in enhancing SCA production is in progress.

Future Prospects

With the use of the antidigoxin SCA as a model, we demonstrate the feasibility of applying *B. subtilis* as an efficient host to produce and secrete single-chain

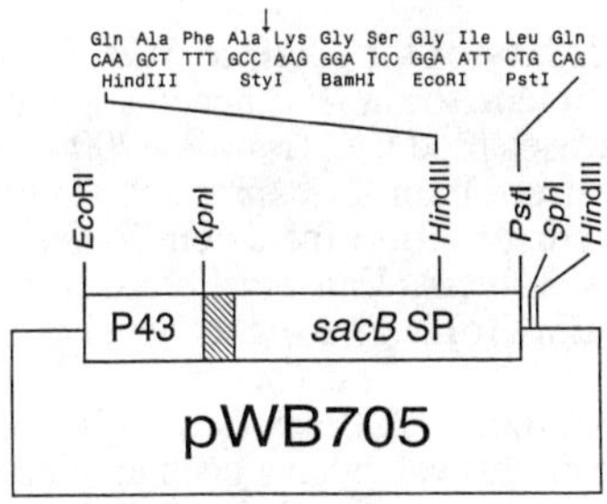

Figure 1. Structural features of the expression-secretion vector pWB705. P43 and *sacB* SP represent the constitutively expressed *B. subtilis* promoter and the modified signal sequence from levansucrase. The hatched area represents the ribosome binding site. Sequences flanking the signal peptidase cleavage site (marked by an arrow) are illustrated.

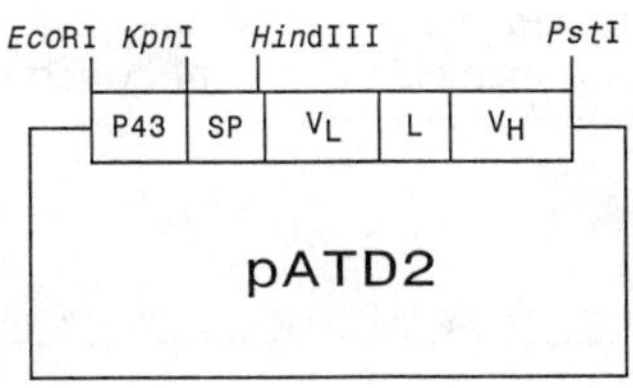

Figure 2. The expression vector, pATD2, for the production of an antidigoxin single-chain antibody fragment. V_L and V_H represent the sequences encoding the variable domains of the light and heavy chains of 26-10. L is the sequence encoding a 15 amino acid linker which links the two variable domains together. P43 and SP are the promoter and signal peptide, respectively.

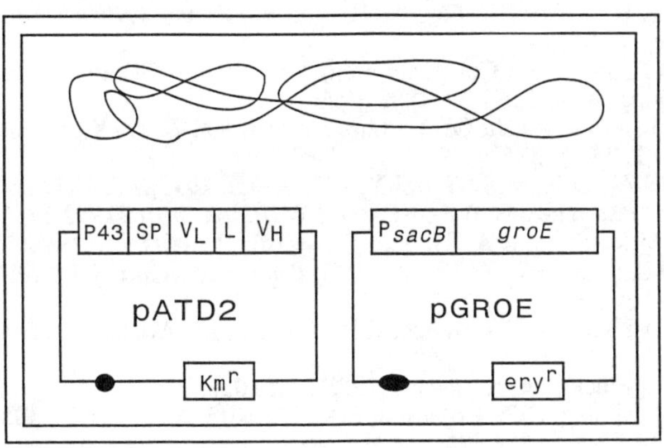

Figure 3. A binary vector system for the enhanced production of the antidigoxin single-chain fragment. pGROE is a pE194(cop6) derivative which carries the *B. subtilis groE* operon regulated by the sucrose-inducible *sacB* promoter. The pUB110 and pE194(cop6) replicons are symbolized a closed circle and an closed ellipse, respectively. Km^r and ery^r indicate the kanamycin and erythromycin resistance markers, respectively.

antibody fragments. Instability of the secreted SCA fragments can be minimized by using the protease deficient strain. Further characterization of the remaining residual extracellular proteases produced from WB700 may allow the development of an even better protease-free strain for expression studies. With the observation that antidigoxin SCA precursors form inclusion bodies, there are many possible strategies to overcome this problem. Enhanced production of antidigoxin SCAs via the co-expression of GroES/GroEL at a higher level is one of the approaches to partially overcome this problem. The use of other molecular chaperones and components involved in secretion (22) may help to increase the production yield. Various fermentation techniques which have been applied successfully in E. *coli* to increase both the cell density and the production yield of the product (*47-48*) can potentially be applied to the *B. subtilis* system. It would be exciting to develop B. subitlis to produce high quality antibody derivatives at a high level in a low production cost.

Acknowledgments

This research is supported by a research contract from the Bristol-Myers Squibb Pharmaceutical Research Institute (to S.L. Wong) and a NSERC strategic grant (S.L. Wong). S.L. Wong is a medical scholar of Alberta Heritage Foundation for Medical Research.

Literature Cited

1. Roberts, S.; Cheetham, J.C.; Rees, A.R. *Nature* **1987**,*328*, 731.

2. Huston, J.S.; Levinson, D.; Mudgett-Hunter, M.; Tai, M.-S.; Novotny, J.; Margolies; M.N.; Ridge, R.J.; Bruccoleri, R.E.; Haber, E.; Crea, R.; Oppermann, H. *Proc. Natl. Acad. Sci. USA* **1988**,*85*, 5879.
3. Riechmann, L.; Clark, M.; Waldmann, H.Winter, G. *Nature* **1988**,*332*, 323.
4. Queen, C.; Schneider, W.P.; Selick, H.E.; Payne, P.W.; Landolf, N.F.; Duncan, J.F.; Avdalovic, N.M.; Levitt, M.; Junghans, R.P.Waldmann, T.A. *Proc. Natl. Acad. Sci. USA* **1989**,*86*, 10029.
5. Brinkmann, U.; Reiter, Y.; Jung, S.-H.; Lee, B.; Pastan, I. *Proc. Natl. Acad. Sci. USA* **1993**,*90*, 7538.
6. Das, C.; Kulkarni, P.V.; Constantinescu, A.; Antich, P.; Blattner, F.R.; Tucker, P.W. *Proc. Natl. Acad. Sci. USA* **1992**,*89*, 9749.
7. Xiang, J.; Koropatnick, J.; Qi, Y.; Luo, X.; Moyana, T.; Li, K.; Chen, Y. *Immunology* **1993**,*78*, 574.
8. Kostelny, S.A.; Cole, M.S.; Tso, J.Y. *J. Immunol.* **1992**,*148*, 1547.
9. Huyghues-Despointes, B.M.P.; Nelson, J.W. *Biochemistry* **1992**,*31*, 1476.
10. Haynes, M.R.; Stura, E.A.; Hilvert, D.; Wilson, I.A. *Science* **1994**,*263*, 646.
11. Lesley, S.A.; Patten, P.A.; Schultz, P.G. *Proc. Natl. Acad. Sci. USA* **1993**,*90*, 1160.
12. Boss, M.A.; Kenten, J.H.; Wood, C.R.; Emtage, J.S. *Nucleic Acids Res.* **1984**,*12*, 3791.
13. Skerra, A.; Pluckthun, A. *Science* **1988**,*240*, 1038.
14. Better, M.; Chang, C.P.; Robinson, R.R.; Horwitz, A.H. *Science* **1988**,*240*, 1041.
15. Wulfing, C.; Pluckthun, A. *Mol. Microbiol.* **1994**,*12*, 685.
16. Skerra, A; .Pluckthun, A. *Protein Eng.* **1991**,*4*, 971.
17. Pantoliano, M.W.; Bird, R.E.; Johnson, S.; Asel, E.D.; Dodd, S.W.; Wood, J.F.; Hardman, K.D. *Biochemistry* **1991**,*30*, 10117.

18. Huston, J.S.; Meredith-Hunter, M.; Tai, M.-S.; McCarney, J.; Warren, F.; Haber, E.; Oppermann, H. *Methods Enzymol.* **1991**,*203*, 46.
19. Harwood, C.R. *Trends Biotech.* **1992**,*10*, 247.
20. Chang, S. *Methods Enzymol.* **1987**,*153*, 507.
21. Behnke, D. In *Biology of Bacilli: Applications to industry*; Doi, R.H., McGloughlin, M., Eds.;Butterworth-Heinemann: MA,1992, pp 143-188.
22. Nagarajan, V. In *Bacillus subtilis and other gram-positive bacteria*; Sonenshein, A.L., Hoch, J.A., Losick, R., Eds.;American Society for Microbiology: Washington, D.C.,1993, pp 713-726.
23. Ferrari, E., Jarnagin, A.S., Schmidt, B.F. In *Bacillus subtilis and other gram-positive bacteria*; Sonenshein, A.L., Hoch, J.A., Losick, R., Eds.;American Society for Microbiology: Washington,D.C.,1993, pp 917-937.
24. Wang, L.-F., Doi, R.H. In *Biology of Bacilli: Applications to Industry*; Doi, R.H., McGloughlin, M., Eds.;Butterworth-Heinemann: Stoneham, MA,1992, pp 63-104.
25. Ogasawara, N. *Gene* **1985**,*40*, 145.
26. Arbige, M.V., Bulthuis, B.A., Schultz, J., Crabb, D. In *Bacillus subtilis and other gram-positive bacteria*; Sonenshein, A.L., Hoch, J.A., Losick, R., Eds.;American Society for Microbiology: Washington,D.C.,1993, pp 871-895.
27. He, X.-S.; Bruckner, R.; Doi, R.H. *Res. Microbiol.* **1991**,*142*, 797.
28. Pero, J., Sloma, A. In ; Sonenshein, A.L., Hoch, J.A., Losick, R., Eds.;American Society for Microbiology: Washington, D.C.,1993, pp 939-952.
29. He, X.-S.; Shyu, Y.-T.; Nathoo, S.; Wong, S.-L.; Doi, R.H. *Ann. NY Acad. Sci.* **1991**,*646*, 69.
30. Wu, X.-C.; Lee, W.; Tran, L.; Wong, S.-L. *J. Bacteriol.* **1991**,*173*, 4952.
31. Isaacs, J.D.; Watts, R.A.; Hazleman, B.L.; Hale, G.; Keogan, M.T.; Cobbold, S.P.; Waldmann, H. *Lancet* **1992**,*340*, 748.
32. Wong, S.-L.; Ye, R.; Nathoo, S. *Appl. Environ. Microbiol.* **1994**,*60*, 517.
33. Wang, P.-Z.; Doi, R.H. *J. Biol. Chem.* **1984**,*259*, 8619.
34. Wong, S.-L. *Gene* **1989**,*83*, 215.
35. Wu, X.-C.; Ng, S.-C.; Near, R.I.; Wong, S.-L. *Bio/Technology* **1993**,*11*, 71.
36. Steinmetz, M.; LeCoq, D.; Aymerich, S.; Gonzy-Treboul, G.Gay, P. *Mol. Gen. Genet.* **1985**,*200*, 220.
37. Mudgett-Hunter, M.; Margolies, M.N.; Ju, A.; Haber, E. *J. Immunol.* **1982**,*129*, 1165.
38. Pantoliano, M.; Alexander, P.; Dodd, S.; Bryan, P.; Rollence, M.; Wood, J.Fahnestock, S. *J. Cell. Biochem.* **1989**,*13A*, 91.
39. Hochuli, E.; Bannwarth, W.; Dobeli, H.; Gentz, R.; Stuber, D. *Bio/Technology* **1988**,*6*, 1321.
40. Skerra, A.; Pfitzinger, I.; Pluckthun, A. *Bio/Technology* **1991**,*9*, 273.
41. Phillips, G.J.; Silhavy, T.J. *Nature* **1990**,*344*, 882.
42. Kusukawa, N.; Yura, T.; Ueguchi, C.; Akiyama, Y.; Ito, K. *EMBO J.* **1989**,*8*, 3517.
43. Lee, S.C.; Olins, P.O. *J. Biol. Chem.* **1992**,*267*, 2849.
44. Schmidt, A.; Schiesswohl, M.; Vîlker, U.; Hecker, M.; Schumann, W. *J. Bacteriol.* **1992**,*174*, 3993.
45. Li, M.; Wong, S.-L. *J. Bacteriol.* **1992**,*174*, 3981.
46. Horinouchi, S.; Weisblum, B. *J. Bacteriol.* **1982**,*150*, 804.
47. Carter, P.; Kelley, R.F.; Rodrigues, M.L.; Snedecor, B.; Covarrubias, M.; Velligan, M.D.; Wong, W.L.T.; Rowland, A.M.; Kotts, C.E.; Carver, M.E.; Yang, M.; Bourell, J.H.; Shepard, H.M.Henner, D. *Bio/Technology* **1992**,*10*, 163.
48. Pack, P.; Kujau, M.; Schroeckh, V.; Knupfer, U.; Wenderoth, R.; Riesenberg, D.; Pluckthun, A. *Bio/Technology* **1994**,

RECEIVED May 17, 1995

Chapter 9

Trichoderma reesei, a Promising Novel Host for Antibody Production

Eini Nyyssönen[1] and Sirkka Keränen[2]

[1]1VTT Biotechnology and Food Research, P.O. Box 1503, FIN-02044 VTT, Finland
[2]Genencor International Inc., 180 Kimball Way, South San Francisco, CA 94080

The exceptional secretion capacity of the filamentous fungus Trichoderma reesei was harnessed for antibody production. T. reesei was able to assemble heterodimeric Fab fragments in active proteins, which were secreted into the culture medium. Reasonable yields were obtained when a Fab fragment was fused to a fungal cellulase, CBHI. The yield of this fusion protein, CBHI-Fab, was 150 mg/l, whereas the Fab fragment as such was produced in only 1 mg/l quantities. Studies on the positive role of the endogenous fusion preceding the antibody heavy chain were carried out. Unexpectedly, the CBHI fragment was shown to act favorably already at the transcriptional level either by stabilizing the messenger or enhancing transcription. At least one limiting step in secretion, which was overcome by the use of CBHI, was protein translocation into the lumen of the endoplasmic reticulum. Furthermore, it was shown that the CBHI-heavy Fd chain was produced in gram per liter range but most of it was cleaved inside the cell, resulting in the CBHI core-linker protein, which was then secreted to the culture medium as efficiently as extracellular T. reesei proteins.

The Trichoderma reesei Host System

Trichoderma reesei. The filamentous fungi Trichoderma spp. are commonly found in soil in all climatic zones. One mesophilic species, Trichoderma reesei, has drawn significant attention for decades from both academia and industry. T. reesei is in fact a single strain, QM6a, which was isolated after the Second World War in The Solomon Islands, northwest of Australia, where it was found to efficiently destroy the U.S. Army cotton tents. This property of cellulose degradation has paved the way for the exploitation of T. reesei in industrial production of hydrolytic enzymes (1).

As a saprophytic fungus T. reesei is able to survive and flourish on complex substrates due to its exceptional ability to secrete enzymes for substrate degradation to readily metabolizable breakdown products. The secretion capacity has been substantially improved for industrial purposes employing classical genetic breeding methods such as mutation and screening (2). Filamentous fungi form tubular structures called hyphae, the tips of which have been shown to be the site for

0097–6156/95/0604–0108$12.00/0

protein synthesis and secretion. The size of the pores on the wall along the filament limits passage of proteins larger than 15-20 kD, and it has been suggested that either the tip wall has larger pores or that the proteins traverse together with nascent wall polymers (3, 4). The hypersecreting T. reesei strains used in industry have been reported to be morphological mutants including strains with thin and branched mycelium (5). A defect in cell wall synthesis can also lead to improved productivity (4) and in one hypercellulolytic T. reesei strain, Rut-C30, high productivity is believed to be related to enhanced amounts of the endoplasmic reticulum (6).

The Cellulase System. T. reesei has hitherto mainly been used in the biotechnical industry for production of cellulases, which are a group of enzymes able to hydrolyze cellulose through synergistic action. Cellulases are being utilized increasingly in various industrial processes such as food and feed processing (7), textile, detergent and pulp and paper industry (2). Thus, the fermentation technology is well developed for T. reesei.

Cellulases can be divided into three major types of enzymes, namely cellobiohydrolases (CBH), endoglucanases (EG), and cellobiase (b-glucosidase). Two genes encoding cellobiohydrolases have been isolated and characterized from T. reesei: cbh1 (8, 9) and cbh2 (10, 11). Four endoglucanases have been isolated up to date: egl1 (12, 13), egl2 (14) and recently, two distinct small endoglucanases, egl3 (15) and egl5 (16). Most of the cellulases exhibit similar structures with a globular core domain containing the hydrolytic active site and a wedge-shaped cellulose binding domain (CBD) (17, 18) connected by an interdomain linker region of extended conformation due to high O-glycosylation. The 3-D structure of the CBHI core domain (19) and an NMR structure of the cellulose binding domain (20) have been reported.

The cellulase production has increased up to 10-fold in the best mutant strains, which have been obtained after several cycles of mutagenesis, when compared with the natural isolate QM6a (5). The highest reported levels of extracellular protein obtained with mutant strains are over 35 g/l (21) and in cellulase-inducing conditions CBHI comprises 60% of the proteins secreted to the culture medium (22).

Filamentous Fungi in Heterologous Protein Production

Transformation methods and expression vectors. Utilization of T. reesei for heterologous protein production was made possible by development of a transformation system (23). In T. reesei transformation the cell walls of hyphae are first enzymatically degraded resulting in osmotically sensitive cells, protoplasts. Transforming DNA is added in the presence of Ca2+ and the cells are induced to fuse by addition of polyethylene glycol (PEG). DNA molecules are apparently internalized during this fusion, since transformants are not obtained without PEG. Transformed protoplasts are plated on an osmotically balanced regeneration medium, with selection based on transformed DNA (24). The fungal selection marker can be linked to the expression vector or to the expression cassette, or it can be transferred to the fungus simultaneously on separate DNA, since cotransformation works efficiently. Autonomously replicating plasmids have not been found for T. reesei, and thus the transforming DNA must integrate into the genome. In contrast to the yeast Saccharomyces cerevisiae, in which non-replicating plasmids almost always integrate into the genome by homologous recombination, the frequency of homologous recombination in filamentous fungi is low (25).

CBHI is a single-copy gene product and is secreted to the culture medium in the grams per liter range under induced conditions, thus cbh1 is believed to be preceded by a strong promoter. To date cbh1 promoter is most widely exploited to drive recombinant protein production in T. reesei, though there is currently increasing interest for finding new promoters active on soluble media, the use of which would facilitate later stages of manufacturing, e.g. in down-stream processing and purification of the product.

Authenticity of the Fungal Made Products. Expansion of the exploitation of T. reesei from a cellulase producer to a heterologous protein producer was justified not only because of the exceptional secretion capacity but also because T. reesei, as well as other filamentous fungi, combine many of the complexities of a eukaryote with the ease of manipulation of a microorganism. Recent reports have confirmed that animals and fungi share a common ancestor, a flagellated protist, and are thus more closely related to animals than plants (26). This relative evolutionary proximity anticipates that fungal hosts may produce more authentic animal proteins than other microbial hosts.

The authenticity of animal proteins produced in the filamentous fungi has not been examined in detail, but to the extent to which it has been investigated the fungal systems appear promising. More data is available on heterologous proteins produced in Aspergillus spp. for which it has been shown that the signal sequences are correctly processed and the recombinant proteins exhibit authentic biological activity. N-Linked glycans are added and the extent of glycosylation and glycan linkage has been shown to be similar to that of the authentic host (27). N-Glycosylation of T. reesei CBHI resembles the high-mannose type N-glycosylation of animal cells (28). If the original protein is post-translationally acetylated at the N-terminus, identical acetylation has been observed in the Aspergillus-made protein. The Aspergillus-system has been shown to allow correct folding and disulfide bond formation. Furthermore, Aspergillus was not only shown to process the pre-sequence of the human tissue plasminogen activator (htPA), but also a pro-sequence which requires an aminopeptidase activity (27, 29, 30). However, whereas the htPA was found to be glycosylated to the authentic extent when expressed under the constitutive A. nidulans triose phosphate isomerase (tpi) promoter in A. nidulans, the same protein in the same species was overglycosylated when production was driven by the A. niger alcohol dehydrogenase (adhA) promoter. The adhA directed expression of human granulocyte colony stimulating factor (HGCSF) and a-1-antitrypsin also resulted in hyperglycosylated proteins. The overglycosylated protein forms were secreted at levels of 1 mg per liter of culture medium, whereas the htPA exhibiting the correct extent of glycosylation yielded only 100 mg per liter. This excessive glycosylation is believed to be due to a stall in protein transport (29, 31).

Yields of Heterologous Proteins Obtained from Filamentous Fungi. Expression of heterologous proteins of fungal origin has been much more successful than production of mammalian proteins in T. reesei, as well as in other filamentous fungi. A high yield of 700 mg/l of the Hormoconis resinae glucoamylase P was obtained when expressed under the strong cbh1 promoter. The yield obtained was 20-fold that obtained from the natural host (32). Examples of recombinant fungal proteins obtained in the grams per liter range from Aspergillus species include glucoamylase, phytase and lipase (33).

Production of mammalian proteins in any of the filamentous fungal species usually results in initially fairly low amounts of the product. Reported yields range from 1 mg/l to 50 mg/l. Examples of therapeutically important proteins produced in filamentous fungi include several human proteins: epidermal growth factor (hEGF), growth hormone (hGH), interleukin (hIL-6), tissue plasminogen activator (htPA), a-1-antitrypsin, granylocyte macrophage colony stimulating factor (hGMCSF) and lactoferrin (29, 30, 34).

However, the initially low levels of a heterologous protein can be significantly improved by combining classical breeding methods and molecular biological techniques. An outstanding example is how the initially very low yield of 14 mg/l of bovine chymosin expressed in Aspergillus niger var. awamori was improved to commercially viable levels of 1.3 g/l of active chymosin (35 - 37). This chymosin production method has been approved by the US Food and Drug Administration (FDA) and the recombinant protein is currently being marketed (Nigel Dunn-Coleman, Genencor International, personal communication, 1993).

Safety Considerations. Approval for manufacturing requires tests to show that the filamentous fungal host is not a mycotoxin producer. Some Trichoderma species are capable of synthesizing a range of mycotoxins, such as trichothecenes, which are associated with a wide variety of human and animal intoxications (38, 39). Production of antiviral, antitumor and anti-weed agents by Trichoderma sp. has also been reported (6, 40). However, at least T. viride loc.sit. and T. polysporum have negligible toxicity to rats: an LD50 in excess of 4 g/kg. Furthermore, cellulase preparation from T. viride loc.sit. has been used as a digestive aid in the preparation of geriatric foods in Japan for several years (6).

Antibody Production by T. reesei

Construction of Strains Producing Antibodies. The exceptional secretory capacity of T. reesei makes it an attractive host for production of heterologous proteins and prompted us to try how well this fungal system would suit for production of antibodies. The model antibody used to test the ability of T. reesei in antibody production was the murine anti-2-phenyloxazolone IgG1 antibody and the following derivatives of it were produced: light chain only, Fab and CBHI-Fab fragments (41; Nyyssönen and Keränen, Curr. Genet., in press) and single chain antibody (unpublished) (Figure 1). A single chain antibody (SCA) had the variable, antigen binding domains of the antibody chains joined by an attaching peptide, which was derived from the linker region of the T. reesei CBHI (42). Fab molecules consist of a complete light chain and an engineered, truncated heavy chain: the heavy Fd chain (VH and CH1). The fusion antibody, CBHI-Fab, is a Fab molecule with the heavy Fd chain fused at its N-terminus to the core-linker region of CBHI (41).

All these derivatives of the murine antibody were produced under the promoter and terminator sequences of the strongly expressed T. reesei cbh1 gene and directed for secretion by the cbh1 signal sequence (41)(Figure 1). SCA was produced also under the constitutive gpd promoter of the filamentous fungus A. nidulans (unpublished)(Figure 1).

The production of Fab and CBHI-Fab required transformation of T. reesei by two separate expression vectors. Thus selection of the antibody chains required two distinct fungal selection markers, one for the light chain and another for the heavy chain expression vector (41). The light chain expression vector was coupled with the phleomycin resistance gene from Streptococcus hindustans (43), flanked by the A. nidulans gpd promoter and trpC terminator regions (44). Both heavy chain expression vectors contained the A. nidulans amdS gene for selection of transformants able to grow on acetamide as the sole carbon source (45). The vectors encoding SCA did not contain a fungal selection marker, but were transferred to T. reesei simultaneously with a plasmid containing the phleomycin resistance gene.

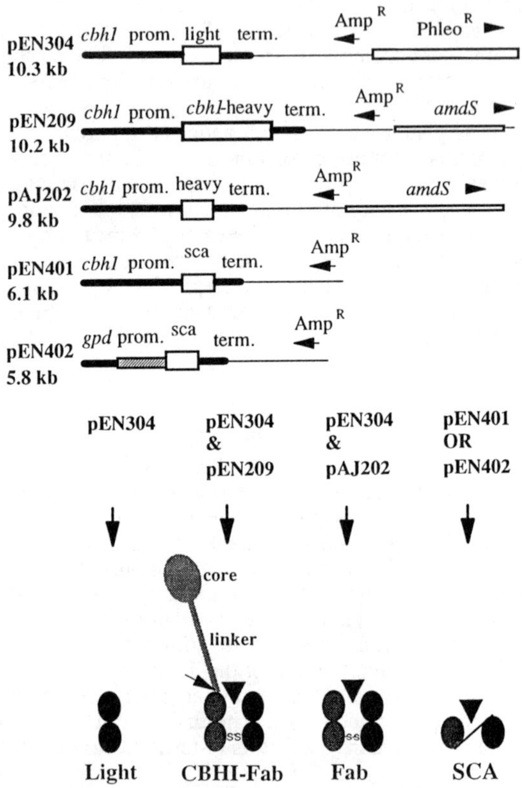

Figure 1. Schematic presentation of the Trichoderma expression vectors encoding the antibody chains and the Trichoderma synthesized antibody fragments. Only the interchain disulfide bonds are presented in the dimeric antibodies. Black triangles depict antigens. The cleavage site of an uncharacterized T. reesei protease is indicated by an arrow on the schematic CBHI-Fab fusion molecule. The black bar connecting the globular domains of a SCA molecule represents the linker peptide derived from the T. reesei CBHI linker.

The hypercellulolytic T. reesei strain, Rut-C30, was the recipient strain for transformations. Transformants obtained were cultivated for four to five days in shake flasks using Solka Floc cellulose or whey as inducing carbon sources, and the antibody production was screened from supernatants by ELISA. For quantitation of production levels the microtiter wells were coated with anti-Ox IgG3 antiserum, and for determination of immunoreactivity with Ox22-BSA (41).

Production Levels of Fungal Synthesized Antibodies. Yields of the light chain only, SCA and Fab were in the range of tens of micrograms to one milligram per liter. The light chain producers were most difficult to obtain, and eventually only two transformants of the total of 130 transformants screened produced the light chain in detectable amounts. These two transformants produced the light chain in similar amounts and the best yields obtained were 0.2 mg/l from a four day shake flask cultivation (41). The best yield of SCA, 1 mg/l, was obtained under the control of the heterologous A. nidulans gpd promoter on glucose medium. In this preliminary shake flask cultivation the yield on glucose medium was 20-fold higher than that from the same strain in the cellulase-inducing conditions. This may not reflect the promoter efficiencies only, since the yield seemed to correlate with pH: the lower the pH, the better the yield. The glucose medium was acidic (pH 3) at the end of the cultivation, in contrast to the final pH 5.5 in cellulase-inducing medium. Nevertheless, in the similar cultivation of the SCA producing strain on glucose and in cellulase-inducing conditions, no SCA was detected when production was driven by the cbh1 promoter (unpublished). The results now obtained in SCA production using the gpd promoter on glucose medium open new prospects for antibody production in well defined, soluble media better suited for large-scale production and down-stream processing.

The best yield was obtained for production of the CBHI-Fab fusion antibody. The yield of secreted, immunologically active CBHI-Fab was 150 mg/l (corresponding to 60 mg/l of Fab fragments) in a four day fermentor cultivation, whereas the yield of Fab fragment obtained in similar cultivation was only 1 mg/l (41). The significant difference in yield between the Fab and CBHI-Fab production was observed throughout the fermentor cultivation and was not due to a more advanced growth state of the CBHI-Fab producer, since the amount of total protein was 1.4 times and total cellulase activity, measured as FPU (filter paper units)(46), was 1.6 times higher in the Fab than in the CBHI-Fab fermentation. The total protein in the CBHI-Fab fermentation was 4.3 g/l, which is much lower than the highest reported protein levels of over 35 g/l obtained with T. reesei (21). Thus it can be expected that the already reasonable yield of 150 mg/l of CBHI-Fab could be further improved by optimizing the cultivation conditions.

Furthermore, it was observed that the T. reesei strains constructed for production of the CBHI-Fab fusion antibody produced large amounts of CBHI core-linker protein originating from the CBHI-heavy Fd fusion protein. The four-day cultivation in a 15 liter fermentor resulted in production of the CBHI core-linker protein in the gram per liter range, CBHI-Fab in the hundred milligrams per liter range and Fab fragments in yet smaller concentrations. Both Fab and the CBHI core-linker protein were proteolytic cleavage products of the fusion protein which thus must have been synthesized rather efficiently (Nyyssönen and Keränen, Curr. Genet., in press).

Purification of the Fusion Antibody, CBHI-Fab. The CBHI-Fab fusion protein was purified from the fermentation growth medium. Crude purification of CBHI-Fab was performed by anion exchange chromatography in which the CBHI-Fab protein exhibited an elution behavior similar to that of the wild type CBHI protein. Fractions containing CBHI-Fab were further purified by affinity chromatography based on antigen binding. The purification resulted in a wide elution peak containing CBHI-Fabs and also Fab-like fragments. Quantitation of the antibodies present in the anion exchange fractions with ELISA showed that only about 0.02% of the fusion protein was cleaved immediately after fermentation. As evidenced by Silver-stained SDS-PAGE, the fusion protein was mainly eluted before the Fab fragments indicating that the CBHI-Fab may have altered antigen binding properties. The last purification step was performed to separate these two antibody forms, CBHI-Fab and Fab fragments, and was achieved by gel filtration (41).

Characterization of the Fusion Antibody and its Cleavage Product. In non-reducing SDS-PAGE the antibody chains from both CBHI-Fab and Fab (the cleavage product) migrated together, in a similar fashion to the idiotypic Fab, indicating proper formation of the interchain disulfide bonds, as was also expected due to activity detected already in the initial screening with ELISA. According to the results of SDS-PAGE the antibody chains did not seem to be aberrantly glycosylated, although it must be borne in mind that the IgG Fab molecules do not naturally contain N-linked glycans (41).

It was not possible to obtain the N-terminal sequence of the light chain or of the CBHI-heavy Fd chain. Both CBHI and the light chain have a glutamyl residue after the signal sequence cleavage site and this seems to have been modified. Pyroglutamyl residue modification could be expected, as it normally occurs in the CBHI protein (47). These data suggest, however, that the signal sequence of these engineered proteins has been correctly processed. Amino acid sequence obtained from the N-terminus of the Fd heavy chain released from the fusion chain revealed that the Trichoderma protease had cleaved the heavy Fd chain from the CBHI-fusion after tyrosine in the CBHI linker region, two amino acid residues before the authentic N-terminus of the heavy Fd chain. At the time the construction of the CBHI-Fab expression vector was made the exact boundaries of the domains (core, linker or cellulose binding domain) were not known and the fusion actually contained part of the cellulose binding domain (CBD) instead of only the core and the extended natural linker region, as intended. The wedge-shaped CBD domain has a hydrophilic flat surface, which has been shown to be important in the binding of the CBD to cellulose. This surface has three tyrosine residues, which have their aromatic side chains positioned parallel to the surface (17, 20). The CBHI-Fab molecule contained only the first of these three tyrosines, after which the cleavage had taken place resulting in the Fab molecule and the CBHI core-linker protein. It is likely that the conformation was altered in the fusion molecule, making this tyrosine-containing region susceptible to proteolysis. This specific cleavage could be mimicked in vitro using chymotrypsin, which is known to cleave after tyrosine. Chymotrypsin is not considered an enzyme of choice for highly specific cleavages, in contrast to Factor Xa, for example, which requires a four amino acid recognition sequence (Ile-Glu-Gly-Arg)(48 - 50). However, the cleavage of CBHI-Fab by chymotrypsin seemed to be very specific. If the structural features of the cleavage site could be applied more generally, chymotrypsin would be an inexpensive choice. Alternatively, the yet uncharacterized T. reesei protease may be exploited for efficient cleavage of fusion proteins.

The hapten-binding affinity of the antibodies was measured by a fluorescence quenching assay (51, 52). This method could not be applied to the fusion molecule due to interfering fluorescence by the CBHI domain, which prevented the determination of the quenching maximum. When the immunoreactivity of the CBHI-Fab fusion was measured on ELISA plates coated with the hapten, the CBHI-Fab was 5-12 times less immunoreactive than the idiotypic Fab, the activity being dependent on protein concentration. This suggests that the CBHI portion of the fusion protein either sterically hindered proper antigen binding or impaired the correct assembly of the immunoglobulin chains in the antigen binding region. However, the fluorescence quenching assay was well applicable to the cleaved Fab fragment, which is of the same size as the control Fab prepared by papain-cleavage from the idiotype antibody. The Fab molecules proteolytically produced from the CBHI-Fab showed idiotypic affinity, i.e. 2 x 106 M-1 in the fluorescence quenching assay and full immunogenic activity in ELISA. These results indicate that T. reesei can be used in the production of authentic antibodies with regard to binding properties.

Positive Effect of the CBHI Fusion in Antibody Production

The working hypothesis for making a gene fusion was to use an endogenous protein as a 'secretion carrier', which would 'pull' the antibody molecule through the secretory pathway. The T. reesei cellulase, CBHI was chosen to be the carrier, since it is the major cellulase secreted into the culture medium in cellulase-inducing conditions. Previously published results had indicated that better yields were obtained by fusing an endogenous protein or most of it to a heterologous protein (36, 53 - 56). The rate-limiting step was believed to be in secretion, leading to intracellular protein accumulation or rapid degradation of the heterologous product. It has been suggested that the resultant fusion protein is either more transport competent or that the endogenous component protects the product from degradation. However, systematic studies on rate-limiting steps in the production of heterologous proteins in filamentous fungi do not exist. Pieces of reported data gathered from different studies may give contradictory results. This suggests that the limitations are case-dependent, affected by the strain, the promoter used and the protein produced.

The production of Fab fragments was increased over 50-fold by fusing the core-linker region of the CBHI to the heavy Fd chain. This result was in accordance with the previously reported data and prompted studies of the beneficial role of the CBHI in antibody production. Better understanding of the limitations should pave the way to targeted improvement of T. reesei production strains for industrial utilization.

Integration of the Antibody Chain Encoding Expression Cassettes. The positive effect of CBHI in antibody production was studied by maintaining the integration of the light chain expression cassette unchanged in strains producing either the light chain only, Fab or CBHI-Fab. This was achieved by using the same light chain-producing strain as a basis for Fab and CBHI-Fab production i.e. the heavy chain expression vectors were transferred to a T. reesei strain, which already had the light chain expression cassette integrated in its genome. Southern results indicated that two copies of the light chain expression cassette had integrated into the cbh1 locus and inactivated the endogeous cbh1 gene (Nyyssönen and Keränen, Curr. Genet., in press). The heavy chain expression cassette in the strains producing Fab or CBHI-Fab had integrated at different locations in the genome.

Methods for Studing the Role of CBHI in Antibody Production. In order to elucidate at which level the CBHI fusion conferred its positive effect in antibody production, detailed quantitation and comparison of both protein and RNA levels were done for all the strains. Endoglucanase I (EGI) was used as an endogenous reference protein, which in cellulase inducing conditions comprises 10% of the total extracellular protein (22). The original transformation host strain, Rut-C30, and the strains producing the light chain, Fab and CBHI-Fab were cultivated in shake flasks and the CBHI-Fab producer was also cultivated in the 15 l fermentor. The cultivations were performed in less concentrated medium than usually in order to obtain an even and more comparable growth between the strains. Due to the dilute medium lower antibody yields were obtained than usually. It was noticeable that in this cultivation, as in the earlier ones, the growth of antibody-producing strains was slower than that of the host strain, Rut-C30 (Nyyssönen and Keränen, Curr. Genet., in press). This altered growth could have been due to the missing CBHI activity required for optimal growth when cellulose is used as carbon source, though it is not excluded that the antibody production itself affected the growth. The secreted and intracellular proteins were analyzed by ELISA and Western analysis with antibodies of several specificities: against CBHI core, CBHI cellulose binding domain, EGI and the light chain. Total RNA was analyzed by Northern and RNA slot blot analyses. Quantities of the mRNA levels were determined measuring the radioactivity from the excised slots by a liquid scintillation counter. Production of the light chain by the strains producing the light chain only, Fab and CBHI-Fab was compared with the production of EGI both with respect to the steady state mRNA levels and the molar amounts of the protein in the culture medium. In the same way the production of CBHI in the host strain Rut-C30, and CBHI core-linker protein in the strain producing the fusion antibody were compared using EGI as a reference. The data obtained with the antibodies were compared with those of the homologous protein EGI in order to estimate the efficiency of antibody production at post-transcriptional level (translation, folding, degradation and secretion)(Nyyssönen and Keränen, Curr. Genet., in press).

Beneficial Effect at Transcriptional Level. Contrary to our working hypothesis the CBHI fragment in the fusion antibody was shown to have a positive effect already at the transcriptional level. In the strain producing CBHI-Fab the steady state mRNA level for the fusion heavy chain was significantly higher than that for the light chain both in the shake flask cultivation and throughout the fermentor cultivation and was detected both by the heavy chain and cbh1 probes in Northern and RNA slot blot analyses. However, the heavy Fd chain mRNA from the Fab producer was seen only after a long exposure, in contrast to the light chain mRNA in the same strain. In order to study whether the integration site affected the heavy chain expression, the shake flask cultivation and RNA slot blot analysis was repeated using four strains, two of which were Fab producers and two CBHI-Fab-producing strains and the same results as above were obtained (Nyyssönen and Keränen, Curr. Genet., in press). Thus it seems that the low levels of the heavy Fd chain mRNA were not due to the integration site.

Our results have shown that one of the limiting steps in antibody production in T. reesei occurred already at the transcriptional level and that it could be circumvented by using an endogenous gene fusion. In all of the three antibody chain constructions the cbh1 controlling regions and the signal sequence were identical. Thus it is unlikely that the high mRNA level of the fusion would have been caused by improved initiation of transcription. More probably the heavy chain mRNA was inherently unstable and the cbh1 part either enhanced transcription or had a stabilizing effect on the mRNA (Nyyssönen and Keränen, Curr. Genet., in press).

Positive Role of CBHI in Secretion. The quantitation results showed that the levels of secreted antibodies were lower than expected on the basis of the mRNA levels. This could have been due to inefficient secretion of the heterologous protein products.

At least one limiting step in secretion, which was overcome by the use of CBHI, was the protein translocation into the lumen of the endoplasmic reticulum or in close association with it. This was suggested, since accumulation of light chains containing the signal sequence was detectable only in the strain producing the light chain only or Fab fragment as such. This result was obtained by Western analysis, in which the intracellular sample of the strain producing the light chain only showed a clearly visible extra band migrating above the light chain. The extra band was less visible in the intracellular sample of the Fab producer and was not detected at all in the strain producing CBHI-Fab. These results suggest that the extra band detectable by the light chain specific antibody could represent light chains still containing the 17 amino acid long signal peptide (Nyyssönen and Keränen, Curr. Genet., in press).

Introduction of the heavy chain to the strain producing the light chain affected the processing of light chains, which suggests that the factors related to the accumulation of unprocessed light chains function in the ER lumen. It is also possible that translated light chains remained in the cytoplasm if the secretory pathway was blocked by previously synthesized, poorly secreted light chains. The latter idea is supported by the fact that folding of the light chains is independent of the heavy chains in B-lymphocytes (57).

The Western analysis from the intracellular protein samples showed also that the efficient cleavage of the fusion heavy chain resulting in large amounts of the CBHI core-linker protein in the culture medium took place already inside the cell. A protein detectable by CBHI-core specific antibody, but not by an antibody specific for the CBHI cellulose binding domain or the heavy Fd chain was seen in Western analysis from both the intracellular sample and the culture medium of the CBHI-Fab-producing strain. It was estimated by ELISA and by end point dilution series in Western analysis that the growth medium from the fermentor cultivation contained 50 mg/l CBHI-Fab and as much as 800 mg/l CBHI core-linker protein, whereas the heavy chains released from the CBHI component were not detected in the culture medium or inside the cell, which could be due to efficient intracellular proteolytic degradation (Nyyssönen and Keränen, Curr. Genet., in press).

Detailed analysis of the protein and mRNA levels of the antibody chains and their comparison to EGI and egl1 mRNA levels showed that production of the light chain only was strongly reduced and was enhanced tenfold by introducing the heavy Fd chain expression cassette to the same strain. Formation of a dimeric molecule thus resulted in more efficient secretion of the light chain. In this strain the production of complex Fab molecules was still inefficient, which may at least partly be explained by the very low levels of the heavy Fd chain mRNA (Nyyssönen and Keränen, Curr. Genet., in press).

Production of the intact CBHI-Fab fusion antibody was more efficient but its secretion was still 2 - 4 fold less efficient compared to the secretion of endogenous fungal protein, EGI. The yields actually obtained were much lower than expected based on the comparison with EGI production which may have been due to the limiting levels of the light chain mRNA and to the slow growth of the CBHI-Fab-producing strain. However, these results suggest that the CBHI fusion also aids secretion of the antibodies, although not to the level of endogenous T. reesei proteins. The cleavage product of the fusion heavy chain, CBHI core-linker protein, was secreted to the culture medium as efficiently as extracellular T. reesei proteins.

Conclusions and Future Prospects

The antibody yields obtained from T. reesei were very promising considering that no strain improvement or optimization of fermentation conditions were made. However, our results showed that production of antibodies was far less efficient than that of the endogenous secreted enzymes.

Yield Improvement. Yields of the light chain only, SCA and Fab were in the range of tens of micrograms to one milligram per liter. This result was in accordance with the previously reported difficulties in harnessing the secretion machinery of filamentous fungi for secretion of heterologous proteins, especially those of mammalian origin (27). The strains producing antibodies grew more slowly on cellulosic substrate, which was most likely due to the lack of CBHI activity, although it was not excluded that the antibody production per se was harmful to the fungus. Nevertheless, optimization of culture conditions and the production strains is expected to overcome the poor growth and thus improve the yields of antibodies.

One requirement for optimal production of a heterodimer, which did not yet materialize in this study, is that the subunits should be produced in stoichiometric amounts. The low mRNA level of the heavy Fd chain seemed to be limiting to the overall yield of Fab molecules. The yields of CBHI-Fab were also still low, although the secretion was no longer strongly impaired. It is likely that in the case of CBHI-Fab production the mRNA levels for the light chain were limiting. Thus optimization of the expression for equal amounts of the antibody chains is also expected to increase the yields. The fusion strategy was clearly beneficial but requires development of an efficient cleavage procedure for removal of the fusion part preferably after secretion into the culture medium. It would also be instructive to study whether active dimeric antibodies could be produced by synthesizing both chains as fusion proteins.

In the preliminary studies the yields of SCA were better when expressed under a constitutive promoter on glucose medium than under the cbh1 promoter in cellulase-inducing conditions. Soluble media would be advantageous in down-stream processing and purification procedures in large-scale production of antibodies. However, complex and robust media have generally been used for induction of the cbh1 promoter. Furthermore, under cellulase-inducing conditions T. reesei secretes large amounts of extracellular fungal proteins, which hamper and complicate purification procedures, in contrast to T. reesei grown on glucose medium. Therefore it is interesting to see how the recently isolated T. reesei promoters active on glucose (58) would perform in antibody production.

The reasonable yield of 150 mg/l of the CBHI-Fab fusion antibody was obtained without any attempts to optimize the cultivations, instead the medium and cultivation conditions optimized for cellulase production were directly applied. Furthermore, no attempts to improve the production strain were made. Against this background, the yield of 150 mg/l CBHI-Fab, corresponding to 60 mg/l Fab fragments, showed that the T. reesei system is competitive with other antibody production methods using other microorganisms or hybridoma cells.

Efficient production of heterologous proteins in filamentous fungi has usually been obtained by a random approach, i.e. mutagenesis and subsequent screening for better producers. This approach may also be used to obtain better antibody producers. Combination of molecular biological techniques and classical mutagenesis resulted in A. niger var. awamori industrial production strains producing 1.3 g/l chymosin, when the initial yields obtained from the strain were as low as 14 mg/l (37).

Better Understanding of Molecular Biology of Filamentous Fungi. Ultimately for improved applicability of T. reesei as a production host for antibodies, it would be desirable to circumvent the use of the fusion. This approach would require a better understanding of the rate-limiting steps in heterologous protein production. Studies of the roles of CBHI in enhanced antibody production were carried out in order to obtain data on the rate-limiting steps in heterologous protein production. Further studies need to be carried out, since the results obtained in this study depicted only a steady-state situation, and thus it would be instructive to obtain turnover data in order to further specify the role of CBHI in mRNA stability versus transcription efficiency and to estimate the rate of protein degradation in the case of homologous and heterologous proteins.

The presence of unprocessed antibody chains, observed in this study, suggests that the host proteins required for translocation, folding, assembly, secretion or glycosylation do not efficiently recognize their heterologous targets. Expression of heterologous proteins in fungi may have to go one step further by concomitant expression of proteins of the same heterologous origin involved in these contributory functions. One alternative to such concomitant expression could be to clone the heavy chain binding protein BiP/GRP78 along with the antibody chains. In addition to modulating protein folding and assembly in the ER lumen BiP may be directly involved in the translocation of precursors across the ER membrane. Furthermore it is known that BiP or other 'chaperones' are not always interchangeable, which would support the idea of concurrent expression of BiP from the same heterologous origin as the antibody of interest. This might require altering the BiP retention signal to be suitable for T. reesei (59, 60).

Applications. The promising results obtained with this system anticipate that T. reesei will find application in the industrial production of antibodies. Optimally the antibody products are first refined in E. coli, since this rapid prokaryotic system is excellent for antibody engineering, both in the search for new specificities and in the improvement and alteration of current ones. The eukaryotic microorganism T. reesei has an exceptional secretion capacity, which can undoubtedly be harnessed for antibody production. Furthermore, large scale production and downstream processing technology for T. reesei are already available and the low costs of its growth medium make it clearly competitive with other antibody production systems.

The potential high yields determine the applications. Examples of applications with high consumption of antibodies are in large-scale affinity purification, as catalytic antibodies and in in vivo diagnosis and therapy. As was shown, T. reesei was able to make antibodies with authentic immunoreactivity, an important requirement for affinity and catalytic purposes. For in vivo applications, if full length antibodies are used, further studies are required to determine whether N-glycosylation of T. reesei, which has been shown to resemble high-mannose type glycosylation in animal cells (28), is sufficient to mediate effector functions: cellular cytotoxicity and activation of complement. Furthermore, glycosylation affects protein stability and aberrant glycosylation triggers the immune response non-specifically (61). It is likely that the terminal glycans in T. reesei-made antibodies differ from the glycan structures in the native hosts. However, identical glycosylation to native hosts may be attained by in vitro modifications or by genetically modifying T. reesei.

Literature Cited

1. El Gogary, S.; Leite, A.; Crivellaro, O.; El Dorry, H.; Eveleigh, D.E. In: Trichoderma reesei cellulases: Biochemistry, Genetics, Physiology and Application; Kubicek C.P; Eveleigh, D.E.; Esterbauer, H.; Steiner, W.; Kubicek-Pranz, E.M., Eds.; The Royal Society of Chemistry: Cambridge, UK, 1990, pp. 200-211.

2. Nevalainen, H.; Harkki, A.; Penttilä, M.; Saloheimo, M.; Teeri, T.; Knowles, J. In: Biotechnology in Pulp and Paper Manufacture. Applications and

Fundamental Investigations; Kirk, T.K; Chang, H.-M., Eds.; Butterworth-Heinemann: Boston, MA, 1990, pp. 593-599.
3. Rambosek, J.A.; Leach, J. C.R.C. Crit. Rev. Biotechnol. 1987, 6, 357-393.
4. Wösten, H.A.B.; Moukha, S.M.; Sietsma, J.H.; Wessels, J.G.H. J. Gen. Microbiol. 1991, 137, 2017-2023.
5. Nevalainen, H. Genetic improvement of enzyme production in industrially important fungal strains; Technical Research Centre of Finland, Publications 26.: Espoo, Finland, 1985, 58 p. + app. 34 p.
6. Eveleigh, D.E. In: Biology of Industrial Microorganisms; Demain, A.L.; Solomon, N.A., Eds.; Benjamin/Cummings Publishing Co., Advanced Book Program: Menlo Park, CA, 1985, pp. 487-509.
7. Penttilä, M.; Teeri, T.T.; Nevalainen, H.; Knowles, J.K.C. In: Applied Molecular Genetics of Fungi; Peberdy, J.F.; Caten, C.E.; Ogden, J.E.; Bennett, J.W., Eds.; Cambridge University Press: Cambridge, UK, 1991, pp. 85-102.
8. Shoemaker, S.; Schweickart, V.; Ladner, M.; Gelfand, D.; Kwok, S.; Myambo, K.; Innis, M. Bio/Technology 1983, 1, 691-696.
9. Teeri, T.T.; Salovuori, I.; Knowles, J. Bio/Technology 1983, 1, 696-699.
10. Chen, C.M.; Grizali, M.; Stafford, W. Bio/Technology 1987, 5, 274-278.
11. Teeri, T.T.; Lehtovaara, P.; Kauppinen, S.; Salovuori, I.; Knowles, J. Gene 1987, 51, 43-52.
12. Penttilä, M.; Lehtovaara, P.; Nevalainen, H.; Bhikhabhai, R.; Knowles, J. Gene 1986, 45, 253-263.
13. van Arsdell, J.N.; Kwok, S.; Schweikart, V.L.; Ladner, M.B.; Gelfand, D.H.; Innis, M.A. Bio/Technology 1987, 5, 60-64.
14. Saloheimo, M.; Lehtovaara, P.; Penttilä, M.; Teeri, T.T.; Ståhlberg, J.; Johansson, G.; Petterson, G.; Claeyssens, M.; Tomme, P.; Knowles, J. Gene 1988, 63, 11-21.
15. Ward. M.; Weiss, G.; Larenas, E.; Rey, M.; Clarkson, K. In: Programme and Abstracts: 2nd International Symposium on Trichoderma cellulases and other hydrolases (TRICEL 93). Kirkkonummi, Finland, 1993, A26.
16. Saloheimo, A.; Henrissat, B.; Hoffren A.M.; Teleman, O.; Penttilä, M. Mol. Microbiol. 1994, 13, 219-228.
17. Claeyssens, M.; Tomme, P. In: Trichoderma reesei cellulases: Biochemistry, Genetics, Physiology and Application; Kubicek, C.P; Eveleigh, D.E.; Esterbauer, H.; Steiner, W.; Kubicek-Pranz, E.M., Eds.; The Royal Society of Chemistry: Cambridge, UK, 1990; pp. 1-11.
18. Abuja, P.M.; Schmuck, M.; Pilz, I.; Tomme, P.; Claeyssens, M.; Esterbauer, H. Eur. Biophys. J. 1988, 15, 339-342.
19. Divne, C.; Stahlberg, J.; Reinikainen, T.; Ruohonen, L.; Petterson, G.; Knowles, J.K.C.; Teeri, T.T.; Jones, T.A. Science 1994, 265, 524-528.
20. Kraulis, P.J.; Clore, G.M.; Nilges, M.; Jones, T.A.; Pettersson, G.; Knowles, J.; Gronenborn, A.M. Biochemistry 1989, 28, 7241-7257.
21. Durand, H.; Clanet, M.; Tiraby, G. Enzyme Microb. Technol. 1988, 10, 341-345.
22. Grizali, M.; Brown, R.D.Jr. Adv. Chem. Ser. 1979, 181, 237-260.
23. Penttilä, M.; Nevalainen, H.; Rättö, M.; Salminen E.; Knowles, J.K.C. Gene 1987, 61, 155-164.
24. Timberlake, W.E.; Marshall, M.A. Science 1989, 244, 1313-1317.
25. May, G. In: Applied Molecular Genetics of Filamentous Fungi; Kinghorn, J.R.; Turner, G., Eds.; Blackie Academic & Professional: Glasgow, UK, 1992, pp. 1-25.
26. Wainright, P.O.; Hinkle, G.; Sogin, M.L.; Stickel, S.K. Science 1993, 260, 340-342.

27. Gwynne, D.I. In: Applied Molecular Genetics of Filamentous Fungi; Kinghorn, J.R.; Turner, G., Eds.; Blackie Academic & Professional: Glasgow, UK, 1992, pp. 132-151.
28. Salovuori, I.; Makarow, M.; Rauvala, H.; Knowles, J.; Kääriäinen, L. Bio/Technology 1987, 5, 152-156.
29. Upshall, A.; Kumar, A.A.; Bailey, M.C.; Parker, M.D.; Favreau, M.A.; Lewison, K.P.; Joseph, M.L.; Maraganore, J.M.; McKnight, G.L. Bio/Technology 1987, 5, 1301-1304.
30. Ward, P.P.; Lo, J.-Y.; Duke, M.; May, G.S.; Headon, D.R.; Conneely, O.M. Bio/Technology 1992, 10, 784-789.
31. Saunders, G.; Picknett, T.M.; Tuite M.F.; Ward, M. TIBTECH 1989, 7, 283-287.
32. Joutsjoki, V.V.; Torkkeli, T.T.; Nevalainen, K.M.H. Curr. Genet. 1994, 24, 223-228.
33. Jeenes, D.J.; MacKenzie, D.A.; Roberts, I.N.; Archer, D.B. Biotechnology and Genetic Engineering Reviews 1991, 9, 327-367.
34. Gwynne, D.I.; Buxton, F.P.; Williams S.A.; Garven, S.; Davies, R.W. Bio/Technology 1987, 5, 713-719.
35. Berka, R.M.; Bayliss, F.T.; Bloebaum, P.; Cullen, D.; Dunn-Coleman, N.S.; Kodama, K.H.; Hayenga, K.J.; Hitzeman, R.A.; Lamsa, M.H.; Przetak, M.M.; Rey, M.W.; Wilson, L.J.; Ward, M. In: Applications of Enzyme Biotechnology; Kelly, J.W.; Baldwin, T.O, Eds.; Plenum Press: New York, NY, 1991; pp. 273-292.
36. Ward, M.; Wilson L.J.; Kodama, K.H.; Rey, M.W.; Berka, R.M. Bio/Technology 1990, 8, 435-440.
37. Dunn-Coleman, N.S.; Bloebaum, P.; Berka, R.M.; Bodie, E.; Robinson, N.; Armstrong, G.; Ward, M.; Przetak, M.; Carter, G.L.; LaCost, R.; Wilson, L.J.; Kodama, K.H.; Baliu, E.F.; Bower, B.; Lamsa, M.; Heinsohn, H. Bio/Technology 1991, 9, 976-981.
38. Smalley, E.B.; Strong, F.M. In: Mycotoxins; Purchase, I.F.H., Ed.; Elsevier Scientific Publishing Company: Amsterdam, The Netherlands, 1974, pp. 199-228.
39. Ueno, Y. In: Mycotoxins in Human and Animal Health; Rodricks, J.V.; Hesseltine, C.W.; Mehlman, M.A., Eds.; Pathotox Publishers, Inc.: Park Forest South, IL, 1977, pp. 189-208.
40. Leigh, C.; Taylor, A. In: Mycotoxins and Other Fungal Related Food Problems; Rodricks, J.V., Ed.; American Chemical Society: Washington, D.C., 1976, pp. 228-275.
41. Nyyssönen, E.; Penttilä, M.; Harkki, A.; Saloheimo A.; Knowles, J.K.C.; Keränen, S. Bio/Technology 1993, 11, 591-595.
42. Takkinen, K.; Laukkanen, M.-L.; Sizmann, D.; Alfthan, K.; Immonen, T.; Vanne, L.; Kaartinen, M.; Knowles, J.K.C.; Teeri, T.T. Protein Eng. 1991, 4, 837-841.
43. Drocourt, D.; Calmels, T.P.G.; Reynes, J.P.; Baron, M.; Tiraby, G. Nucleic Acids Res. 1990, 18, 4009.
44. Mattern, I.E.; Punt, P.J.; Unkles, S.; Pouwels, P.H.; van den Hondel, C.A.M.J.J. In: Abstracts of the 19th Lunteren Lectures on Molecular Genetics of Yeasts and Filamentous Fungi and its Impact on Biotechnology. Lunteren, The Netherlands, 1987, pp. 34.
45. Kelly, J.M.; Hynes, M.J. EMBO J. 1985, 4, 475-479.
46. IUPAC (International Union of Pure and Applied Chemistry) Pure and Appl. Chem. 1987, 59, 257-268.
47. Fägerstam, L.G.; Petterson, L.G.; Engström, J. FEBS Lett. 1984, 167, 309-315.

48. Nagai, K.; Thogersen, H.C. Nature 1984, 309, 810-812.
49. Sieg, K.; Kun, J.; Pohl, I.; Scherf, A.; Muller-Hill, B. Gene 1989, 75, 261-270.
50. Haley, P.E.; Doyle, M.F.; Mann, K.G. J. Biol. Chem. 1989, 264, 16303-16310.
51. Methods in Medical Research; Eisen, H.N., Ed.; Year Book Medical Publishers: Chicago, IL, 1964, Vol. X., pp. 115-121.
52. Weir, D.M. In: Handbook of Experimental Immunology, 3rd edition; Blackwell Scientific Publications: Oxford, UK, 1978, ch. 18.
53. Harkki, A.; Uusitalo, J.; Bailey, M.; Penttilä, M.; Knowles, J.K.C. Bio/Technology 1989, 7, 596-603.
54. Contreras, R.; Carrez, D.; Kinghorn, J.R.; van den Hondel, C.A.M.J.J.; Fiers, W. Bio/Technology 1991, 9, 378-381.
55. Baron, M.; Tiraby, G. J. Biotechnol. 1992, 24, 253-266.
56. Jeenes, D.J.; Marczinke, B.; MacKenzie, D.A.; Archer, D.B. FEMS Microbiol. Lett. 1994, 107, 267-271.
57. Tartakoff, A.; Vassalli, P. J. Cell Biol. 1979, 83, 284-299.
58. Nakari, T.; Alatalo, E.; Penttilä, M. Gene 1993, 136, 313-318.
59. Gething, M.-J.; Sambrook, J. Nature 1992, 355, 33-45.
60. Pelham, H.R.B. TIBS 1990, 15, 483-486.
61. Goochee, C.F.; Monica, T. Bio/Technology 1990, 8, 421-425.

RECEIVED May 17, 1995

Chapter 10

Applying Antibody Catalysis to Organic Synthesis

Brian J. Lavey and Kim D. Janda

Departments of Molecular Biology and Chemistry, The Scripps Research Institute, 10666 North Torrey Pines Road, La Jolla, CA 92037

Since the initial reports in 1986 *(1, 2)*, catalytic antibody technology has offered the possibility of designing highly specific catalysts for use in organic synthesis. This chapter will examine the advances in catalytic antibody technology that now allow it to be applied to preparative scale organic chemistry in terms of the mechanics of carrying out antibody reactions for synthesis, the scope of reaction types to which antibody catalysis has been applied, and the possibilities inherent in generating antibodies without using hybridoma techniques.

Preparative Scale Antibody Reactions

In attempting to design experimental conditions for antibody or enzymatic catalysis, a number of chemical engineering problems associated with the optimal use of an expensive biocatalyst must be overcome. These include: determining the optimal times to expose the substrates to the catalyst and preventing product inhibition, as well as separating the catalyst from reactants for the repetition of the reaction cycle. Mutual compatibility of the antibody and the reactants must also be a concern. The conditions used should not denature the antibody and the substrates and products should not decompose in the presence of water. Recently, we have made significant progress in overcoming all of these challenges.

The primary means of dealing with these problems has been to use mixtures of organic solvents and water as the reaction medium for antibody catalyzed reactions. Both micelles *(3)* and simple biphasic systems *(4, 5)* are effective in this regard. Such solvent mixtures alleviate several difficulties associated with biocatalytic reactions. First, they allow the use of relatively nonpolar substrates that are typically of interest in synthetic chemistry. Second, the presence of an organic solvent can help alleviate problems of product inhibition, a problem that is frequently associated, in aqueous environments, with reactions in which the hydrophobic effect plays a large role in the binding of substrates and products. Having immiscible mixtures of solvents also provides an effective means of separating the water soluble antibodies from hydrophobic products when reuse of the catalyst is desired, a technique that has been used extensively in separating water soluble transition metal catalysts from their products and substrates *(6)*. This process has also been automated *(7)* such that a solution of substrate in an apolar organic solvent is pumped into a reaction vessel containing an aqueous antibody solution. The vessel is then shaken for a predetermined period of time based on previously determined reaction kinetics. Once the reaction is

0097–6156/95/0604–0123$12.00/0

complete, the organic solvent, now containing the product, is pumped into a containment vessel for later evaporation and a fresh charge of substrate solution is added to the reaction flask. This process is then repeated under microprocessor control for as many cycles as may be desired without further intervention by the operator prior to workup. For reactions in which a wholly aqueous environment is preferred, the antibody is enclosed by a dialysis bag which allows substrates and products to diffuse through the membrane, but retains the protein *(8)*. This procedure however, is usually associated with a decrease in the rate of product formation, probably due to slow diffusion of the reactants across the membrane.

Applying Antibody Catalysis to Reactions of Synthetic Interest

Among the singular advantages of using antibodies as catalysts is the tremendous selectivity and specificity of the immune system in binding substrates. This allows the designer of a catalytic antibody system to reasonably expect a high degree of regioselectivity and enantioselectivity in the reaction of interest as well as the possibility of carrying out normally disfavored transformations. In this regard, the achievement of an anti-Baldwin ring closure*(9)*, an exo-Diels-Alder reaction *(10)* and a syn elimination reaction *(11)* should be mentioned.

In planning the design of haptens to induce antibody catalysts, some thought should be given to the requirements for catalytic activity, as well as some of the limitations of current knowledge regarding the generation of catalytic functional groups. First, it is crucial to have some knowledge of the mechanism of the reaction of interest and to be able to synthesize a reasonable transition state analog to act as a hapten. If more than just stabilization of a particular geometry is desired, the use of a "bait and switch" strategy *(12-13)* will allow the generation of catalytically active groups in the antibody active site. Current technology provides fairly reliable prospects for inducing catalytically active acidic and basic residues *(13)*. Nucleophilic residue on the other hand have, so far, proven more difficult, as has the combination of multiple active site residues capable of acting in catalytic concert to bring about energetically demanding transformations.

Hydrolysis and Acyl Transfer Reactions. Hydrolytic and acyl transfer enzymes have been applied widely in organic synthesis, and their use has been extensively documented *(14-19)*. Recurring themes include the enantioselective cleavage of meso substrates to give chiral products, kinetic resolution of racemic materials, and the use of enzymes as mild deprotecting reagents *(19)*. Hydrolytic antibodies developed for synthetic purposes have followed these trends and can be expected to do so in the future. In one of our own examples of the synthesis of a chiral material from a meso substrate, *cis*-1,3-diacetoxycyclopent-2-ene was hydrolyzed in chiral fashion to give **2**, a precursor for the synthesis of prostaglandin F_{2a} and its congeners *(20)*. Other applications of hydrolytic antibodies which have been used for synthetic purposes include hydrolytic kinetic resolutions of racemic esters *(21-22)* and the enantioselective formation of lactones *(23)*.

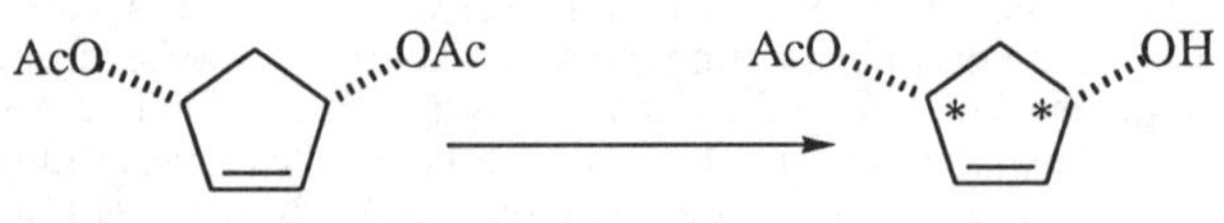

1 **2**

86% ee

Antibodies are also capable of effecting transesterification reactions in aqueous solutions *(24-26)*. In one example, we showed that vinyl ester **3** could be converted to a variety of aryl esters **4** *(24)*. The antibody which effected this conversion was raised against phosphonate ester **5** and used sophisticated ping-pong and induced fit mechanisms, previously thought to belong only in the realm of highly evolved enzymes. In more recent work by the group of Schultz, antibodies have been used to catalyze the formation of peptide bonds *(27)*. As catalysts which can operate easily in an aqueous environment, and do not require the use of protecting groups, antibodies offer interesting, though still distant, possibilities for the synthesis of peptides via fragment coupling procedures.

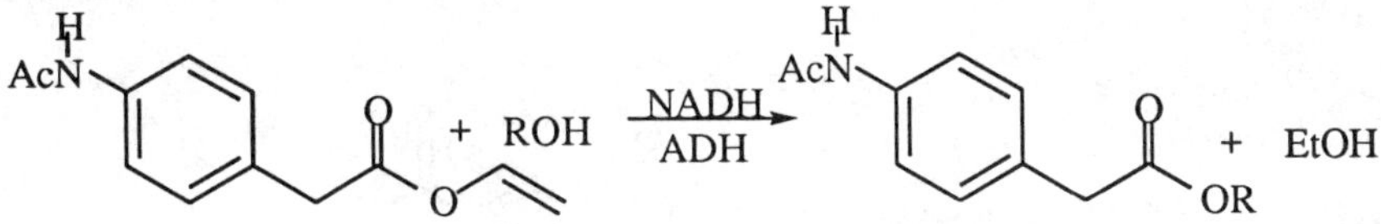

3 **4**

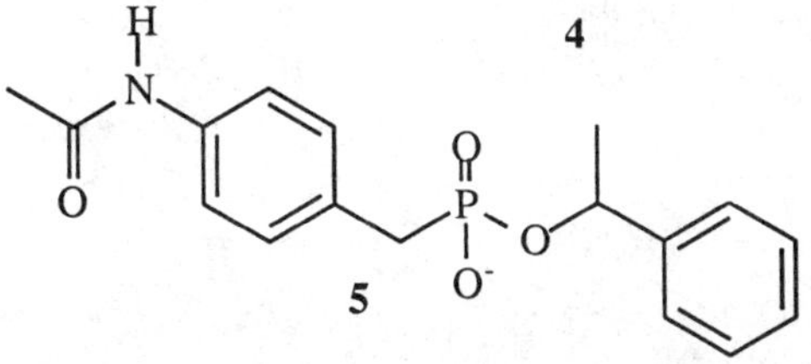

5

Redox Chemistry. In the field of redox chemistry, catalytic antibodies have demonstrated the remarkable specificity and selectivity of the immune system and also diverged significantly from the practices used in enzymatic redox synthesis. Unlike many enzymatic redox reactions, antibody catalysts have been combined with commonly available inorganic oxidizing and reducing agents, thus avoiding the need to obtain and regenerate biological cofactors, a process that is frequently the most difficult aspect of enzymatic redox synthesis.

In the first example of using antibody catalysts in combination with a stoichiometric inorganic reactant, Schultz described the use of sodium cyanoborohydride with antibodies raised against phosphonate **6** to reduce the alpha-keto amide **7** to the corresponding amido alcohol **8** in 99% ee *(28)*. It should also be noted that formation of the opposite epimer is somewhat preferred in the background reaction. This work is noteworthy not just because of the use of $NaBH_3CN$ as an inorganic cofactor, but also because this particular borohydride is a very mild reagent which ordinarily reacts with ketones only very slowly. The use of antibodies to increase the reactivity of ketones toward reduction was followed up by using an antibody to specifically activate a particular ketone for reduction in the presence of another*(29)*. The conversion of **9** to **10** occurs in 96% ee and with greater than 75:1 regioselectivity, a ratio which further demonstrates the utility of the principle of activating substrates toward reagents with which they are ordinarily unreactive. Schultz has also demonstrated that $NaBH_3CN$ does not cause any appreciable degradation of the antibody in greater than 25 turnovers.

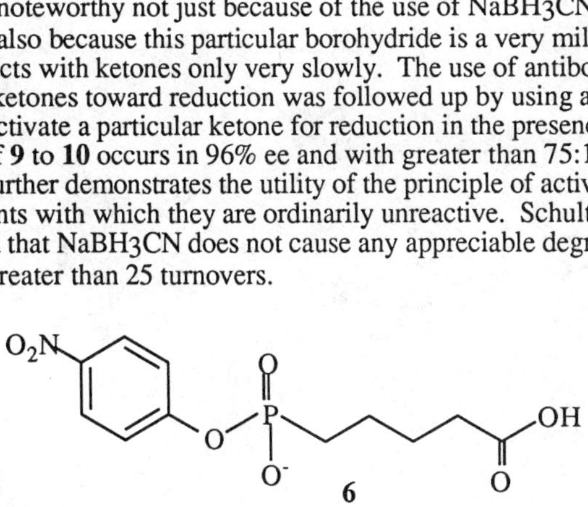

6 Hapten

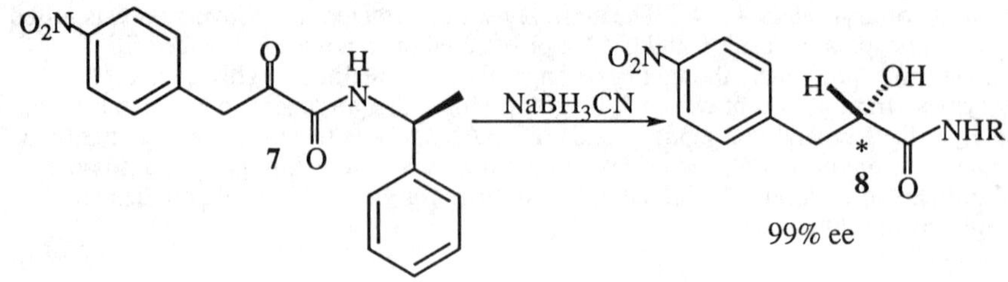

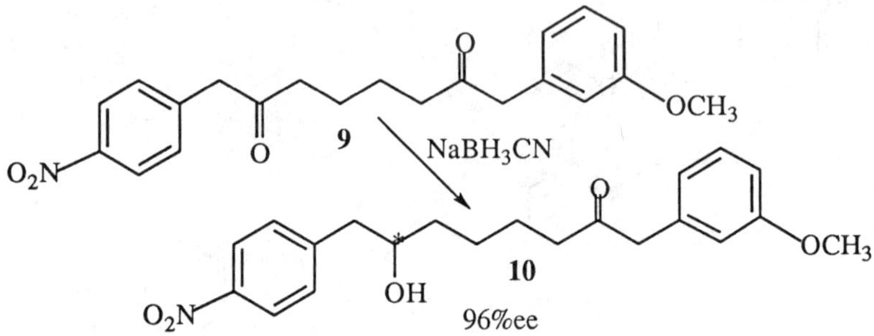

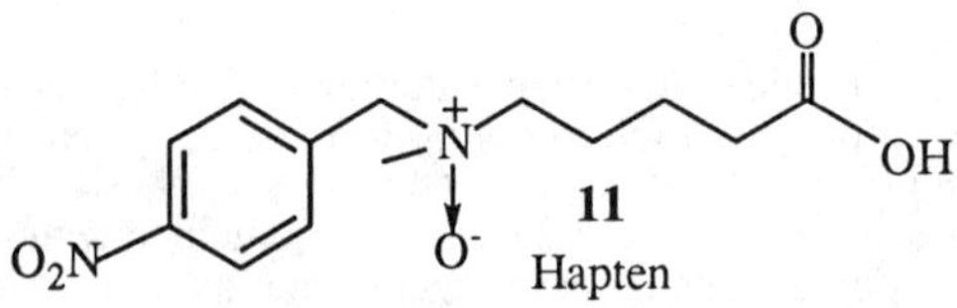

In another case in which a stoichiometric reagent was used successfully in conjunction with antibody activation of substrates, Lerner has reported the enantioselective epoxidation of olefins using Payne's reagent, a mixture of dilute aqueous hydrogen peroxide and acetonitrile *(30)*. In a different oxidation study, directed at the enantioselective synthesis of sulfoxides, sodium periodate has also been used successfully in the presence of antibodies as well *(31)*.

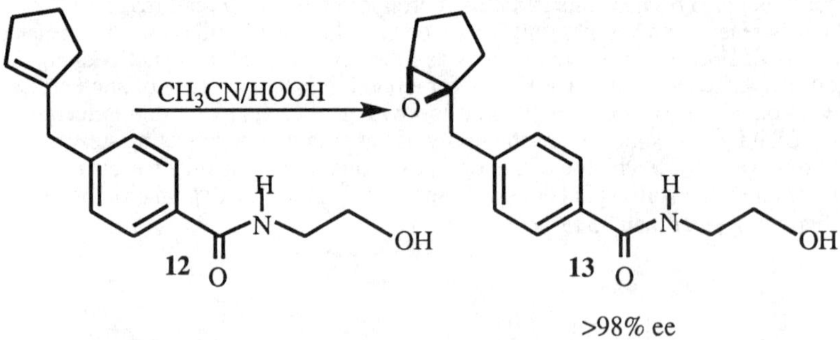

Antibody Mediated Cyclization Reactions. Catalysts which can effect reactions in a highly regio and stereoselective fashion are potentially extremely useful. Biologically significant cyclic natural products which are potentially accessible by antibody catalyzed cyclization reactions include steroids, terpenes, prostaglandins, polyethers, and various macrocycles. Nature, as usual, has led the way in synthesizing a dazzling array of complex skeletons with high stereospecificity, however, reproducing some of the cyclization reactions which occur within these biosynthetic pathways is often very difficult. Most notable among these challenges has been the attempt to mimic the full tetracyclization reaction of squalene oxide to lanosterol as performed by oxidosqualene cyclase *(32-33)*. As might be expected, the requirements for each type of cyclization vary somewhat and therefore, they will be examined separately.

In attempting to carry out carbon-carbon bond forming cyclizations in aqueous media, the chemist is confronted by the need to stabilize both transient carbocations and substrate conformations such that cyclization is faster than attack by water, and also faster than loss of a proton to form an olefin. In addition, the material which is to mediate this transformation must not itself be alkylated by the carbocation intermediates. Finally, according to the Stork-Eschenmoser postulate *(34-35),* the geometry in which the acyclic substrate exists during the cyclization reaction will determine the stereochemistry of the cyclic product. Accordingly, while great strides have been made in using cyclization reactions in synthesis, attempts to carry out cationic cyclization reactions in the laboratory often lead to complex mixtures of products, regardless of whether or not the reaction occurs in acid, water, or an organic solvent.

In the first example of an antibody catalyzed cationic cyclization, we demonstrated that antibodies raised against hapten **14** could convert compound **15** almost exclusively to **16** with only traces of cyclohexene formed as the lone side product *(36)*. In contrast, exposure of **15** to aqueous acidic conditions caused immediate loss of the phenyl dimethyl silane. No **16** was formed under the solvolysis conditions. The clean formation of **16** is significant for a number of reasons. It demonstrates that an antibody can bind small molecules in the very tightly defined geometry necessary for a cationic cyclization reaction to occur. Formation of the silyl alcohol exclusively with the *trans* stereochemistry is also indicative of rigidly defined stereochemistry within the antibody pocket, even to the point of allowing the intramolecular addition of water to compete with the intramolecular elimination of silicon or a proton to form a cyclohexenyl product. The stereochemical definition and lack of side products bode well for the use of antibody mediated cyclization reactions with more complex substrates. The rigid definition of acyclic substrate conformation has also been observed in heteroatomic cyclizations as well.

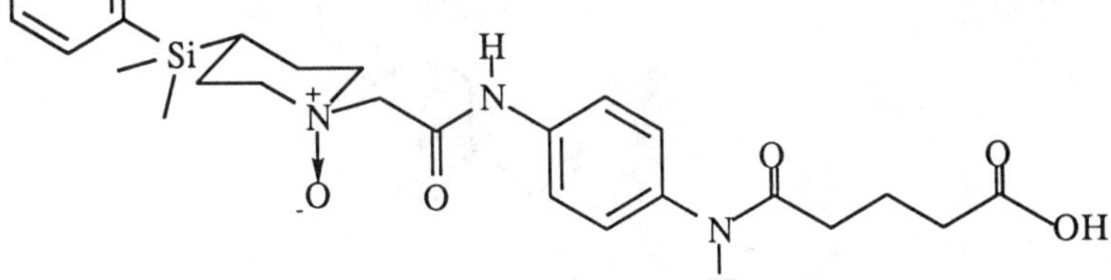

Hapten **14**

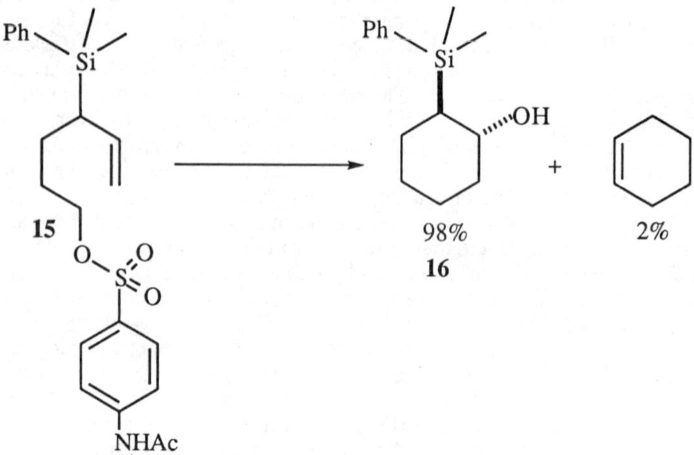

We also reported the first heteroatom based antibody catalyzed cyclization *(9)*, in which N-oxide **17** induced antibodies that transformed epoxy alcohol **18** into tetrahydropyran **19** despite the fact that formation of the tetrahydrofuran **20** is the kinetically preferred reaction according to Baldwin's rules. The difference in energy between the transition states that lead to these two products has been calculated to be approximately 1.8 kcal/mol under aqueous conditions, an amount which should afford a 96:4 ratio of **20** to **19** *(37)*. This difference in energy is due primarily to the necessity of placing the substrate in a conformation which will satisfy the requirement of backside displacement of the departing leaving group by the attacking nucleophile. Using our bait and switch principle, the antibodies which were induced by **17** should have strategically placed residues in the combining site capable of stabilizing developing carbocationic charge on the 6'-carbon atom. The negatively charged oxygen atom of the N-oxide should induce an amino acid residue in the combining site which can act as a general acid to assist in the epoxide opening. The use of a cyclic hapten structure is to insure that the substrate will be organized into the disfavored six membered ring transition state. We have also carried out a similar reaction on preparative scale with compound **21** to give 957 mg of **22** in enantioselective fashion, with resolved starting material as the only side product *(7)*.

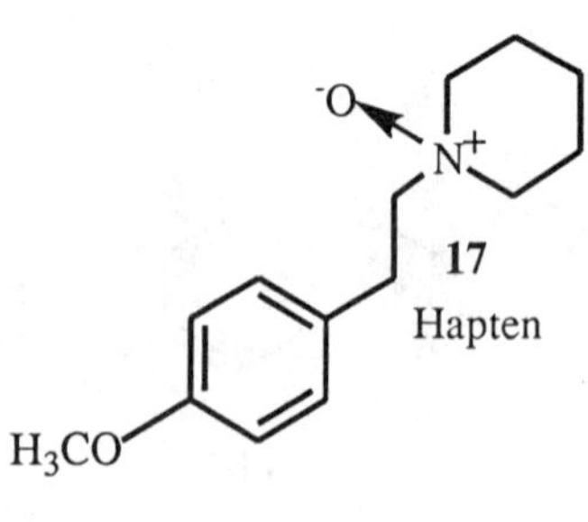

17
Hapten

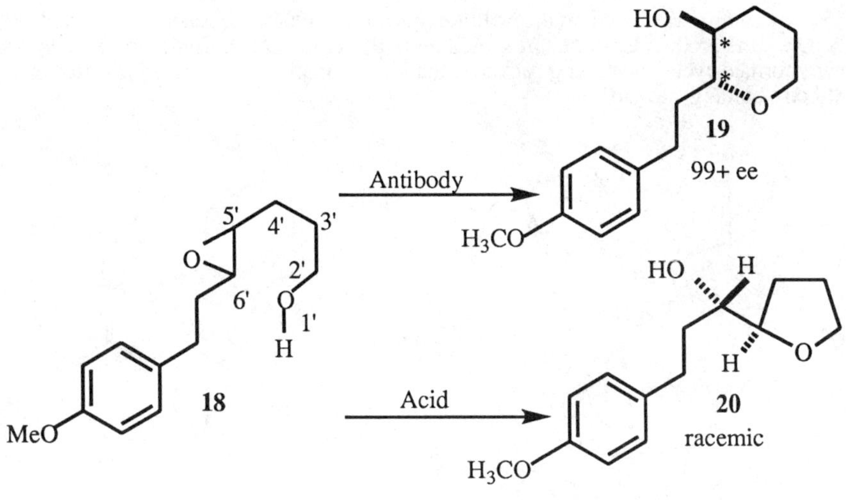

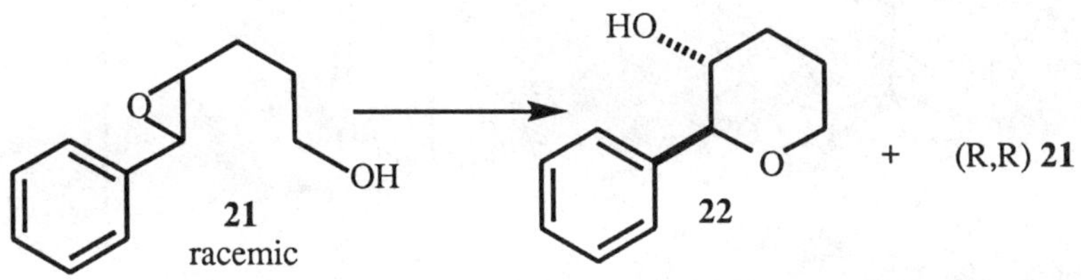

Diels-Alder Reactions. Antibody catalyzed Diels-Alder reactions have been carried out by the groups of Schultz *(38)*, Hilvert *(39)*, Suckling *(40)*, and ourselves *(10)*. As reactions which are enthalpically "downhill" and in which transition states closely resemble products, Diels-Alder transformations are excellent models with which to examine the possibilities of catalysis by approximation. In this vein, the haptens used to induce "Diels-Alderases" have primarily been intended to act as "entropic traps," with no attempt made so far to activate the dieneophile by protonation of an associated carbonyl group as has been done with Lewis acids in organic solvents. Given the reported advantages of using water as a solvent for some Diels-Alder reactions, the application of antibody catalysis to this reaction, with the ensuing possibilities of regio-, diasterio-, and enantiocontrol offers exciting possibilities. Because many Diels-Alder substrates are achiral, while the products frequently have multiple stereocenters, the Diels-Alder reaction allows the dramatic amplification of the chiral atmosphere of the antibody binding site. Given the venerable history of the reaction in synthesis, many interesting applications of an enantioselective Diels-Alder reaction might be cited, including the synthesis of prostaglandins via the Corey lactone *(41)* and the synthesis of sugars via the Danishefsky diene procedures *(42)*.

To date, the most useful hapten design for Diels-Alder reaction antibody catalysts has been put forth by the Schultz group *(38)*. A bridged tricyclic hapten was used to induce a Diels-Alderase antibody for the conversion of **23** and **24** to **25** *(38)*. In this case, it was thought that hapten **26** would provide an appropriate binding site for the reaction without generating a catalyst in which product inhibition would be severe.

This expectation was based on the boatlike transition state for the reaction, depicted as **27**. Once the product is formed, it should relax from the boat configuration in the newly formed cyclohexane ring back to a chair conformation initiating release from the antibody binding site *(38)*.

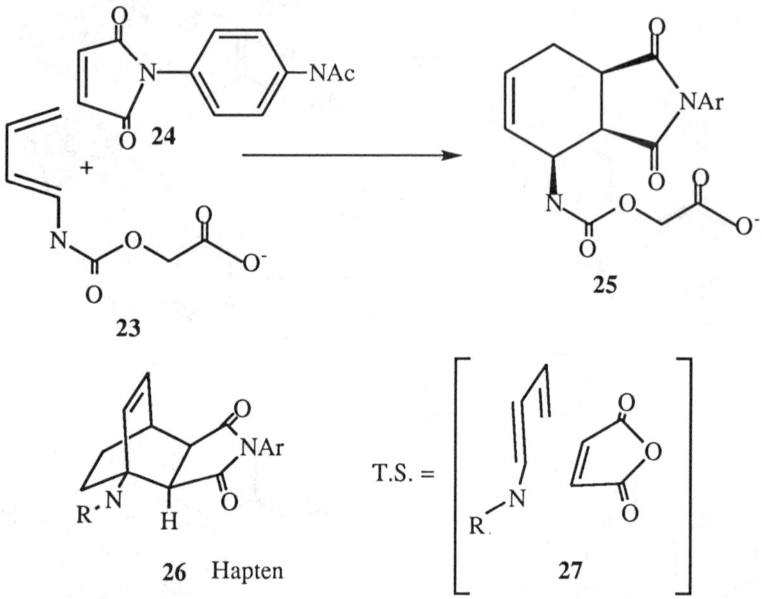

In contrast to the use of transition state analogs as haptens, Suckling has reported that the Diels Alder formation **31** and **32** can be catalyzed by antibodies raised against hapten **33**, which closely resembles the products *(40)*. Such a strategy of hapten design might be expected to have inherent problems associated with product inhibition. In the case of a Diels-Alder reaction, however, only antibodies capable of binding the substrates into a boatlike geometry will be catalysts. As a result, the same mode of product release after relaxation into a chairlike conformation might be expected. How much effect the change in alkyl group has on product release is difficult to judge. That both the N-ethyl and the N-benzyl maleimides were substrates for the reaction is suggestive that this area of the molecule is not critical for binding. Regardless of the particulars of the manner in which this reaction's hapten, substrates, and products are bound, the success of a strategy of immunizing with a hapten resembling the reaction product may have interesting implications for future work.

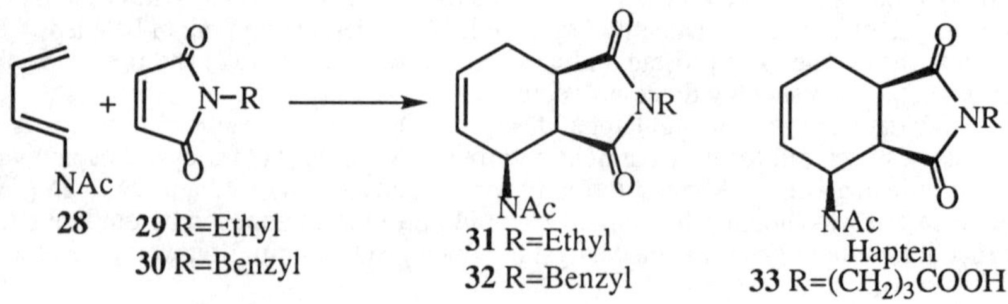

In another application of bicyclooctene haptens for the generation of antibody catalysts, we have demonstrated that it is possible to control the formation of *exo* and *endo* products with complete specificity *(10)*. Mice were immunized with compounds **40** and **41** giving antibodies which bind identical substrates in different conformations to give different products. Specifically, antibodies from mice immunized with **40** converted N-(1-butadienyl)-carbamic acid (**34**) and N,N-dimethyl acrylamide (**35**) to the *endo* Diels-Alder product **36**, while antibodies which formed in response to **41** gave the disfavored *exo* product. In each case, the reactions were completely selective for one enantiomer. Calculations show that formation of the **36** is favored by 1.9 kcal /mol over the *exo* product while the uncatalyzed reaction provides an 85:15 ratio of the *endo* to *exo* products *(10)*. This reaction, in combination with the "Anti-Baldwin" cyclization *(9)* further demonstrates the possibilities inherent in having an antibody enforce rigid kinetic control over a reaction, even to the point of providing products inaccessible to "normal" solution chemistry.

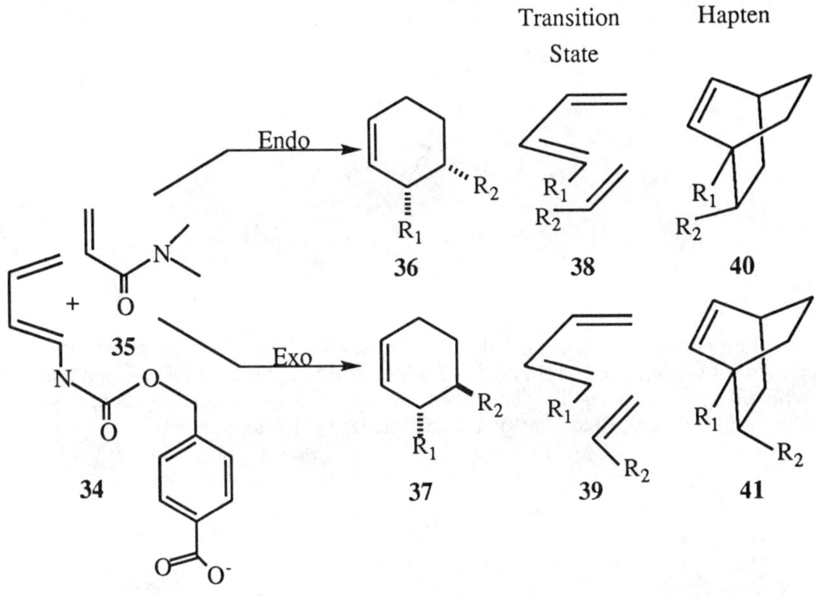

Sigmatropic Rearrangements. Antibody catalysis has been applied to two sigmatropic rearrangements, the Claisen *(43, 44)* and the oxy-Cope *(45)*. Like the Diels-Alder reaction, these concerted reactions are susceptible to catalysis primarily via antibody mediated organization of the substrates into a reactive geometry. Also like the Diels-Alder reaction, these rearrangements should be amenable to the controlled partitioning of the substrates into favored and disfavored reaction pathways based on preferences for chair or boatlike transition states enforced by the antibody binding site and to the catalyst induced placement of new chiral centers in the products of achiral substrates. In the case of the oxy-Cope reaction, one might also expect some acceleration of the reaction if a base were present to polarize the OH bond of the substrate hydroxyl group.

In 1988, the groups of Schultz *(43)* and Hilvert *(44)* each reported the antibody catalyzed conversion of chorismate **42** into prephenate **44**, an important step in the biosynthesis of aromatic amino acids in plants and bacteria. In each case, compound **45** was used as the hapten in an attempt to mimic the presumed geometry of the transition state.

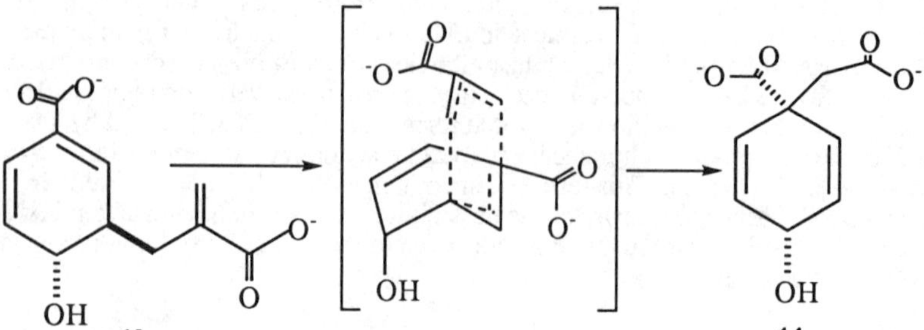

42 **43** **44**

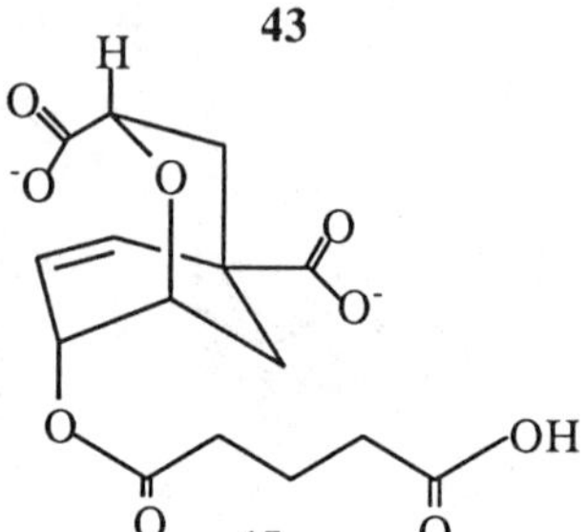

45

In more recent work, carried out by the Schultz group *(45)* the secondary alcohol **46** was converted to aldehyde **47** via an antibody mediated oxy-Cope reaction in which K_{cat}/K_{uncat}=5300. The hapten used was the cyclohexyl compound **48**, which mimics the chairlike transition state typical of Cope rearrangements.

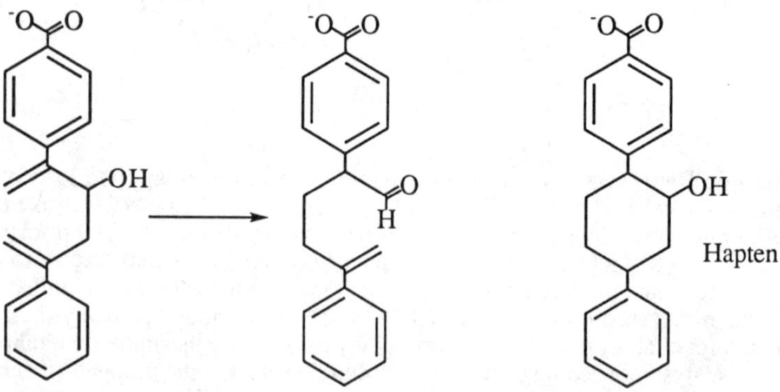

46 **47** **48**

Phage Display Systems

While hybridoma methodology has been used to generate almost all the catalytic antibodies obtained to date, it suffers from a number of limitations. In a system where the immune response to an antigen of interest is on the order of 10^5-10^6 different secreted proteins, hybridoma techniques will allow only a very small fraction to be assayed for catalysis. During the immortalization process, approximately one percent of the available antibody producing B-cells fuse with the carcinoma cell line and not all of the fusions will form viable hybridomas. Of those immortalized cells which survive, the time, labor, and expense of generating sufficient quantities of antibody for kinetic assay allow only a smaller subset of these to be examined. Typically, less than one hundred of the M_{ab}s which bind the hapten most strongly are examined. In some cases, this may in turn cause lower rates of catalytic activity since extremely tight binding of the substrate is detrimental to high reaction rates and turnover numbers. The process of cloning the immune response into *E. coli (46-48)* has provided an alternative method of obtaining antibodies which allows a far higher number of potential catalysts to be surveyed. It also allows the production of antibody catalysts using *in vitro* techniques more likely to be in the sphere of expertise of chemists and biochemists than hybridoma methods.

Generating Phage Display Libraries. In the formation of an antibody library, DNA encoding antibody heavy and light chain genes is obtained either from the spleen of an immunized animal *(49, 50),* or produced semisynthetically *(51).* DNA from either the heavy or the light chains is fused with DNA encoding the carboxyl terminus of gene III of bacteriophage λ *(50, 52).* Elsewhere in the vector, DNA from the alternate chain and a β-lactamase gene are also added. In wild type phage, gene III encodes product coat protein III (cp III), which is expressed on the phage surface and is essential for infection of bacteria. The phage containing the new genes are then used to infect male *E. coli.* The *E. coli* are grown in a medium containing carbenicillin, thereby preventing the growth of bacteria which do not have the vector genes. The resulting infected *E. coli* can synthesize the heavy and light chain proteins. After translation, the chains are transported to the periplasmic space where they fold into a F_{ab}-cp III fusion protein. The *E. coli* are then infected with helper phage which contain genes coding for the proteins needed for phage packaging, including wild type cp III. Competition between the two forms of cp III results in the packaging of phagemids which generally carry only one copy of the F_{ab}-cpIII and several copies of the native cp III needed for future *E. coli* infection. Monovalent F_{ab} display allows the selection of high affinity antibody binders and avoids difficulties associated with chelation by several antibodies at once *(52).* Phage libraries obtained by this process may vary in size from 10^6-10^8 different clones, depending on the source and diversity of the original F_{ab} DNA *(50, 51).*

In a process known as panning, a phage library is screened for binding to a hapten of interest which has been immobilized on a solid support *(52).* The nonbinding phage are then washed away with water and buffer. Bound phage are eluted with acid or dissolved hapten. The binders are used to infect bacteria for further amplification and antibody production. This process is repeated five or six times. When antibodies of catalytic interest are obtained, the phage DNA is isolated, and the cp III gene is removed from the antibody chain gene to which it is attached. The antibody genes may then be converted to plasmids for recloning into other expression vectors for overexpression and further selection for catalytic activity. When compared with hybridoma techniques, the use of antibody phage libraries generated via this method allows several orders of magnitude more antibodies to be surveyed for catalytic activity. In a variation of this process, error prone PCR has been used to afford a form of affinity maturation to provide increased substrate specificity *(53).*

Screening Antibody Libraries with Mechanism Based Inactivators. In designing appropriate means of screening large antibody libraries to obtain catalysts, two seemingly contradictory requirements must be satisfied. In order to differentiate between antibodies which are highly specific for a substrate of interest and those antibodies which exhibit nonspecific binding, fairly rigorous washing conditions are used, ensuring that only the tightest binders are retained. To be an effective catalyst however, an antibody should bind the substrates and products fairly weakly, thereby preventing the energy of activation from being too large and product release from becoming the rate limiting factor. In order to simultaneously achieve both tight binding of the panning hapten and weak binding of the substrates and products, the use of mechanism based inactivators or "suicide substrates" *(54)* for the reaction and reaction mechanism of interest appears to be highly advantageous. Upon binding to the antibody, mechanism based inactivators will form a covalent bond to the protein. When the panning substrates are attached to a solid support via a cleavable linker, the resulting trapped antibody can be released after the washing procedures are complete. In a variation on this theme, certain types of mechanism based inactivators of enzymes can be used to select for a particular type of protein residue or a set of residues.

In the first attempt to make use of covalent binding in panning substrates, our group used **49** to select for cysteine residues via disulfide interchange in the phage bound antibodies *(55)*. After five rounds of panning, the phagemid DNA was isolated, purified, and grown in *E. coli*. The DNA was reisolated, and a sample of the resulting DNA population was sequenced, showing higher levels of codons for unpaired cysteine residues than would occur by chance. One clone was then selected for kinetic investigation. After overexpression in *E. coli*. functional F_{ab} was obtained and shown to be >98% pure by gel electrophoresis. Upon treatment of the antibodies with **50**, thiopyridine was released, thereby indicating nucleophilic displacement by the antibody. The use of radioactive **50** also indicated the formation of a covalent complex between the antibody and substrate. Compounds **51** and **52** were then tested against the F_{ab} to determine if the active site thiol would be catalytically active in hydrolysis reactions. This proved to be the case. The antibody accelerated the reaction rate by a factor of approximately 10^4 and underwent multiple turnovers. The antibody also displayed burst kinetics typical of enzymes which use covalent based catalytic mechanisms. The success of this screening process allows for the direct, premeditated, selection of residues necessary for catalysis. In the future, as we become more sophisticated in the design of haptens and panning substrates, it should provide a means to generate catalysts for more energetically demanding reactions.

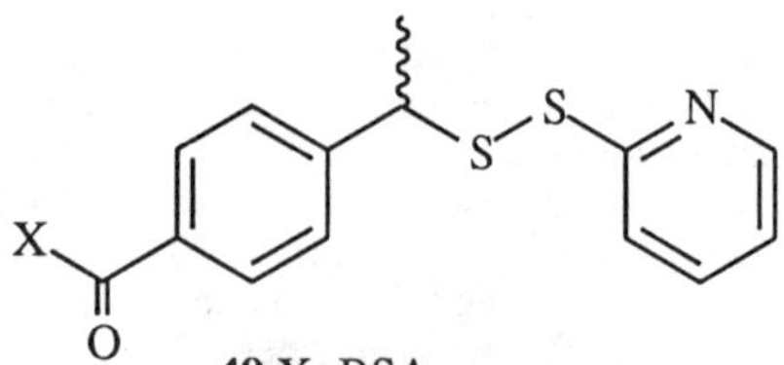

49 X=BSA
50 X=CH_3NH

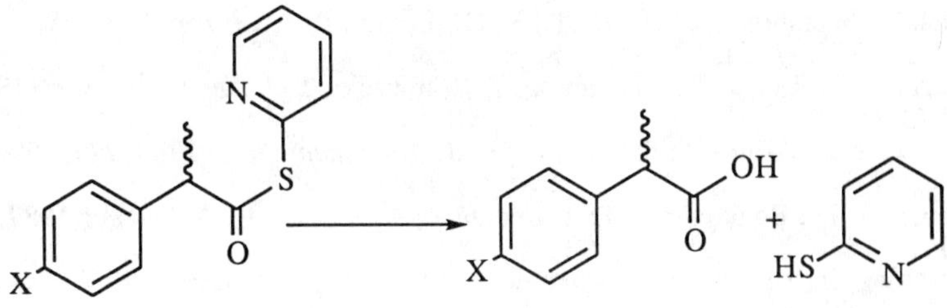

51 X=HCONH
52 X=H

53 X=HCONH
54 X=H

55

Acknowledgments. While we have tried to include as many examples of catalytic antibody reactions that may be applied to synthesis as possible, space limitations have forced us to be selective with the examples used. In addition, oversights and omissions are unavoidable. In those cases where such lapses have occurred we apologize to both the reader and the authors involved. K. D. J. would also like to thank the National Institutes of Health, the National Science Foundation, and the Alfred P. Sloan Foundation who have supported much of the work described here.

Literature Cited.

1 Tramontano, A.; Janda, K. D.; Lerner, R. A. *Science* **1986**, *234*, 1566.
2 Pollack, S. J.; Jacobs, J. W.; Schultz, P. G. *Science* **1986**, *234*, 1570.
3 Durfor, C. N.; Bolin, R. J.; Sugasawara, R.J. and Massey, R.J. *J. Am. Chem. Soc.* **1988**, *110*, 8713-8714.
4 Ashley, J. A.; Janda, K. D. *J. Org. Chem.* **1992**, *57*, 6691-6693.
5 Janda, K.D.; Ashley, J.A.; Jones, T.M.; McLeod, D.A.; Schloeder, D.M.; Weinhouse, M.I. *J. Am. Chem. Soc.* **1990**, *112*, 8886-8888.
6 Hermann, W. A.; Kohlpaintner, C. W. *Angew. Chem. Int. Ed. Engl.* **1993**, 32, 1524-1544.
7 Shevlin, G. S.; Hilton, S.; Janda, K. D. *Bioorganic & Med. Chem. Lett* **1994**, *4*, 297-302.
8 Reymond, J.-L.; Reber, J.-L.; Lerner, R. A. *Angew. Chem. Int. Ed. Engl.* **1994**, 33, 475-477.
9 Janda, K. D.; Shevlin, C. G.; Lerner, R. A. *Science*, **1993**, *259*, 490-492.
10 Gouverneur, V. E.; Houk, K. N.; de Pascual-Teresa, B. ; Beno, B.; Janda, K .D. ; Lerner, R. A. *Science* **1993**, *262*, 204-208.
11 Cravatt, B. F.; Ashley, J. A.; Janda, K. D.; Boger, D. L.; Lerner, R. A. *J. Am. Chem. Soc.* **1994**, *116*, 6013-6014.
12 Janda, K. D.; Weinhouse, M. I.; Schloeder, D. M.; Lerner, R .A.; Benkovic, S. J. *J. Am. Chem. Soc.* **1990**, *112*, 1274-1275.
13 Janda, K. D.; Weinhouse, M. I.; Danon, T.; Pacelli, K. A.; Schloeder, D. M. *J. Am. Chem. Soc.* **1991**, *113*, 5427-5434.
14 Whitesides, G. M.; Wong, C.- H. *Angew. Chem. Int. Ed. Egnl.* **1985**, *24*, 617-718.
15 Faber, K.; Riva, S. *Synthesis* **1992**, 895 and 1049.
16 Drueckhammer, D. G.; Hennen, W. J.; Pederson, R. L.; Barbas, C.F.; Gautheron, C. M.; Krach, T.; Wong, C.- H. *Synthesis* **1991**, 499.
17 Servi, S. *Synthesis* **1990**, 1.
18 Ohno, M.; Otsuka, M. *Organic Reactions* **1989**, *37*, 1.
19 Waldmann, H.; Sebastian, D. *Chem. Rev.* **1994**, *94*, 911-937.

20 Ikeda, S.; Weinhouse, M. I.; Janda, K. D.; Lerner, R. A. *J. Am. Chem. Soc.* **1991**, *113*, 7763-7764.
21 Kitazume, T.; Lin, J. T.; Yamamoto, T.; Yamazaki, T. *J. Am. Chem. Soc.* **1991**, *113*, 8573-8575.
22 Pollack, S. J.; Hsiun, P.; Schultz, P. G. *J. Am. Chem. Soc.* **1989**, *111*, 5961-5962.
23 Napper, A. D.; Benkovic, S. J.; Tramontano, A.; Lerner, R. A. *Science* **1987**, *238*,1041.
24 Wirsching, P.; Ashley, J. A. Benkovic, S. J.; Janda, K. D.; Lerner, R. A. *Science* **1991**, *252*, 680-685.
25 Fernholz, E.; Schloeder, D.; Liu, K. K.- C.; Bradshaw, C. W.; Huang, H.; Janda, K. D.; Lerner, R. A.; Wong, C. -H. *J. Org. Chem.* **1992**, *57*, 4756-4761.
26 Jacobsen, J. R.; Prudent, J. R.; Kochersperger, L.; Yonkovich, S. Schultz, P. G. *Science* **1992**, *256*, 365-367.
27 Jacobsen, J. R.; Schultz, P. G. *Proc. Natl. Acad. Sci. USA* **1994**, *91*, 5888-5892.
28 Nakayama, G. R.; Schultz, P .G. *J. Am. Chem. Soc.* **1992**, *114*, 780-781.
29 Hsieh, L. C.; Yonkovich, S.; Kochersperger, L.; Schultz, P. G. *Science* **1993**, *260*, 337-339.
30 Koch, A.; Reymond, J.- L.; Lerner, R. A. *J. Am. Chem. Soc.* **1994**, *116*, 803-804.
31 Hsieh, L. C.; Stephans, J. C.; Schultz P. G. *J. Am. Chem. Soc.* **1994**, *116*, 2167-2168.
32 Johnson, W. S. *Angew. Chem. Int. Ed. Engl.* **1976**, *15*, 9.
33 Sutherland, J. K. In *Comprehensive Organic Synthesis*; Editor, B. M. Trost; Vol. 3; p. 341.
34 Eschenmoser, A.; Ruzicka, A.; Jeger, O.; Arigoni, D. *Helv. Chim. Act.* **1955**, *226*, 1800-1905.
35 Stork, G.; Burgstahler, A. W. *J. Am. Chem. Soc.* **1955**, *77*, 5068-5077.
36 Li, T.; Janda, K. D.; Ashley, J. A.; Lerner, R. A. *Science*, **1994**, *264*, 1289-1293.
37 Na, J.; Houk, K. N.; Shevlin, C .G.; Janda, K. D.; Lerner, R. A. *J. Am. Chem. Soc.* **1993**, *115*, 8453-8454.
38 Braisted, A. C.; Schultz, P. G. *J. Am. Chem. Soc.* **1990**, *112*, 7430-7431.
39 Hilvert, D.; Hill, K. W.; Nared, K. D.; Auditor, M. -T. M. *J. Am. Chem. Soc.* **1989**, *111*, 9261-9262.
40 Suckling, C. J.; Tedford, M. C; Bence, L. M.; Irvine, J. I.; Stimson, W. H. *Biorg. & Med. Chem. Lett.* **1992**, *2*, 49-52.
41 Corey, E. J.; Weinshenker, N. M.; Schaaf, T. K.; Huber, W. *J. Am. Chem. Soc.* **1969**, *91*, 5675.
42 Danishefsky, S. J.; DeNino, M. P. *Angew. Chem. Int. Ed. Engl.* **1987**, *26*, 15-23.
43 Jackson, D. Y.; Jacobs, J. W.; Sugaswara, R.; Reich, S. H.; Bartlett, P. A.; Schultz, P. G. *J. Am. Chem. Soc.* **1988**, *110*, 4841.
44 Hilvert, D.; Carpenter, S. H.; Nared, K .D.; Auditor, M. -T. M. *Proc. Natl. Acad. Sci. USA* **1988**, *85*, 4953-4955.
45 Braisted, A. C.; Schultz, P. G. *J. Am. Chem. Soc.* **1994**, *116*, 2211-2212.
46 Skerra, A.; Pluckthun, A. *Science* **1988**, *240*, 1038-1041.
47 Orlandi, R.; Gussow, D. H.; Jones, P. T.; Winter, G. *Proc. Natl. Acad. Sci. USA* **1989**, *86*, 3833.
48 Sastry, L.; Alting-Mees, M.; Huse, W. D.; Short, J. M.; Sorge, J.; Hay, B. N. Janda, K. D.; Benkovic, S. J.; Lerner, R. A. *Proc. Natl. Acad. Sci. USA* **1989**, *86*, 5728.
49 Persson, M. A.; Caothien, R. H.; Burton, D. R. *Proc. Natl. Acad. Sci. USA* **1991**, *88*, 2433-2436.

50 Huse, W. D.; Sastry, L.; Iverson, S. A.; Kang, A. S.; Alting-Mees, M.; Burton, D. R.; Benkovic, S. J.; Lerner R. A. *Science* **1989**, *246*, 1275-1281.
51 Barbas III, C. F.; Bain, J. D.; Hoekstra, D. M; Lerner, R. A. *Proc. Natl. Acad. Sci USA* **1992**, *89*, 4457-4461.
52 Barbas III, C. F.; Kang, A. S.; Lerner, R. A.; Benkovic, S. J. *Proc. Natl. Acad. Sci USA* **1991**, *88*, 7978-7982.
53 Gram, H.; Marconi, L. -A.; Barbas III, C. F., Collet, T. A.; Lerner, R. A.; Kang, A. S. *Proc. Natl. Acad. Sci. USA* **1992**, *89*, 3576-3580.
54 Silverman, R. B. *Mechanism-Based Enzyme Inactivation: Chemistry and Enzymology* CRC Press Inc., Boco Raton, FL, 1988;Vol. and Vol. 2.
55 Janda, K. D.; Lo, C. -H. L.; Li, T.; Barbas III, C. F.; Wirsching, P.; Lerner, R. A. *Proc. Natl. Acad. Sci. USA* **1994**, *91*, 2532-2536.

RECEIVED May 17, 1995

Chapter 11

Strong and Specific Peroxidase Activity with an Antibody L Chain–Porphyrin Fe(III) Complex

Tadayuki Imanaka and Masahiro Takagi

Department of Biotechnology, Faculty of Engineering, Osaka University, Yamadaoka, Suita, Osaka 565, Japan

We attempt to generate monoclonal antibodies as host proteins for the functional molecular porphyrin, and thus engineer a new type of catalytic antibody. TCPP (meso-tetrakis (4-carboxyphenyl) porphyin) was chemically synthesized and Balb/c mice were immunized using TCPP as a hapten. Two hybridoma cells (03-1, 13-1), which produce monoclonal antibody against TCPP, were obtained. One of the monoclonal antibodies, Mab 03-1, exhibited enhanced peroxidase activity when TCPP Fe(III) was incorporated. Genes for both H and L chains of monoclonal antibodies were cloned, sequenced and overexpressed using *E. coli* as a host. ELISA and fluorescence quenching method show that the antibody L chains from both Mab 03-1 and Mab 13-1 have specific strong interaction with TCPP. Furthermore, a TCPP Fe(III) complex with the L chain from Mab 13-1 exhibits much higher peroxidase activity than TCPP Fe(III) alone. The enzyme activity was detectable with pyrogallol and ABTS(2,2-azinobis 3-ethylbenzthiazolin-6-sulfonic acid) but not with catechol. This result provides a new concept of catalytic antibody using monoclonal antibody or antibody fragment as a host carrier protein for a functional molecule.

Antibodies have been demonstrated to catalyze a wide variety of transformations such as ester hydrolysis (*1-3*), pericyclic(*4*,) photo-chemical (*5*), ligand substitution(*6*) and redox reactions(*7*.) These new antibodies were designated as catalytic antibodies or abzymes.

Even though several strategies have been reported to generate these catalytic antibodies, we attempt here to engineer a new way to generate a new catalytic antibody with peroxidase activity using a monoclonal antibody as the host carrier protein for a functional molecule, porphyrin (Figure 1). We choose porphyrin because many active centers of various peroxidases and catalases consist of a porphyrin molecule encapsulated by a protein molecule. Previous construction of catalytic antibody used N-methylmeso-porphyrin IX, a presumed transition-state analogue for porphyrin metalation as a hapten. This antibody-porphyrin complex was formed to catalyze the reduction of hydrogen peroxide by several typical

0097–6156/95/0604–0138$12.00/0

chromogenic peroxidae substates(8). We synthesized a TCPP which exhibits higher substrate specificity than N-methylmeso-porphyrin IX. Furthermore, recombinant antibody L chain was found to have specific and strong interaction with porphyrin. The L chain complex exhibited even higher peroxidase activity than the complex of the entire antibody and porphyrin.

Results and Discussion

Raising antibody against porphyrin. TCPP was covalently attached to the carrier protein, keyhole limpet hemocyanin (KLH) using water-soluble carbodiimide, 1-(3-dimethylamino)propyl-3-ethyl-carbodiimide, or carbonyl diimidazol (Figure 2). The conjugates were then purified by column chromatography on Sephadex G-50. Balb/c mice were immunized with the KLH conjugate emulsified in complete Freund's adjuvant. Cell fusion was carried out using p3x63-Ag8.653 myeloma as the fusion partner and polyethylene glycol as the fusion reagent. Two hybridoma cells (03-1, 13-1) were screened and propagated in ascites as described previously(*9*). Two monoclonal antibodies from hybridoma cells (Mab 03-1, Mab 13-1) were purified with a protein A column. Specific interaction of Mab 03-1 and Mab 13-1 with TCPP was examined by ELISA (enzyme-linked immunosorbent assay) and purified monoclonal antibodies were used for further studies.

Gene cloning and expression of monoclonal antibody in *Escherichia coli.*

Hybridoma cells, 03-1 and 13-1, were cultivated and cellular mRNA was purified by Quick Prep mRNA Purification kit (Pharmacia P-L Biochemicals Inc., Milwaukee, WI.). Isolated mRNA was then used to construct cDNA using an oligo(dT)18 primer and reverse transcriptase. Genes for both L and H chains of two monoclonal antibodies (Mab 03-1, Mab 13-1) were amplified using PCR (polymerase chain reaction) and cloned into Immunozap L and H vectors(*10*) respectively. Nucleotide sequences of the cloned DNA fragments encoding L and H chains were determined by dideoxy chain termination method(*11*) (GenBank Acc. No. D29667-D29670). Deduced amino acid sequences were then compared between Mab 03-1 and Mab 13-1 antibodies. The amino acid sequences of the two antibodies were found to be different only in hyper variable regions (CDR 1, 2 and 3). Therefore, Mab 03-1 and Mab 13-1 most likely recognize porphyrin in a different manner.

Induced expression of cloned monoclonal antibody genes was then attempted. However, the expression level of genes cloned into Immunozap vectors was very low. Therefore, genes for the appropriate L and H chain s were subcloned respectively into an expression vector, pET8-c, which has a strong promoter sequence for T7RNA polymerase(*12*.) Overexpression of antibody protein was successfully performed using *E. coli* BL21(DE3) as a host. Antibody proteins could be recovered as inclusion bodies in insoluble cytoplasmic fraction. Inclusion bodies were purified and solubilized using 6 M guanidine hydrochloride. Then, the purified proteins were refolded by dialysis. The homogeneity of purified antibody L and H chains was examined by SDS polyacrylamide gel electrophoresis.

Dissociation constants of monoclonal antibodies

Specific binding between each recombinant antibody chain and porphyrin (TCPP) was examined by ELISA using anti-Mab 03-1 and anti-Mab 13-1 as secondary antibodies. H chains from both monoclonal antibodies did not show antigen binding, while specific binding between independent antibody L chain and TCPP

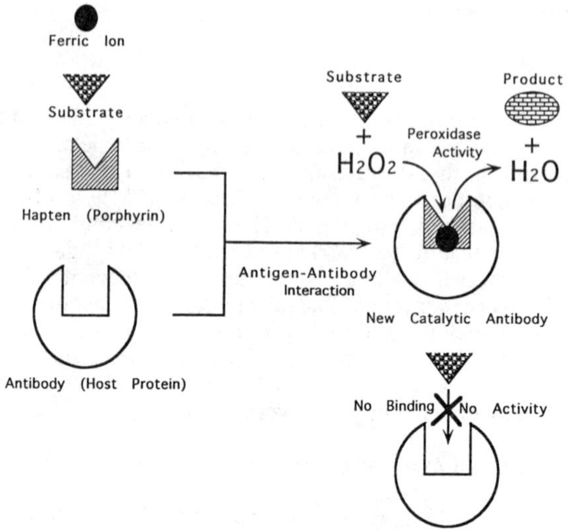

Figure 1. Illustration of a concept for the new catalytic antibody.

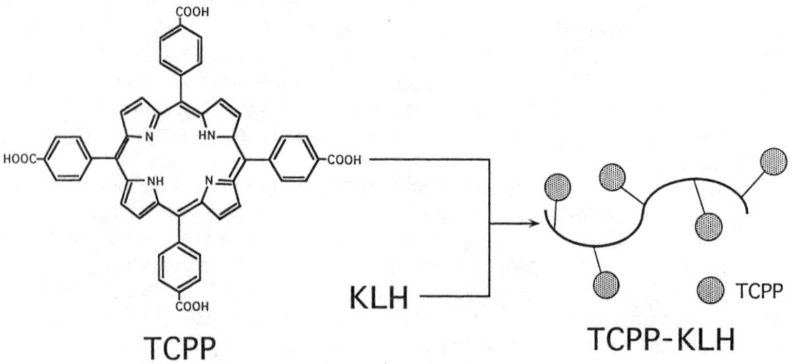

TCPP : meso-tetrakis (4-carboxyphenyl) porphyrin
KLH : keyhole limpet hemocyanin

Figure 2. Structure of TCPP

was observed for both 03-1 and 13-1 antibodies (data not shown). Kinetics of binding was more quantitalively analyzed. Using fluorescence quenching method, dissociation constants (Kd) for Mab 03-1, Mab 13-1 and two L chains were determined(*13*) (Table I). Interestingly, L chains of both 03-1 and 13-1 monoclonal antibodies have specific but slightly weaker interaction with TCPP than Mab 03-1 and Mab 13-1. Moreover, Mab 13-1 did not show any specific interaction against TCPP Fe(III). Independent L chain from Mab 13-1 exhibits specific interaction (Kd=1.4×10^{-5}) with TCPP Fe(III).

Table I: Dissociation constants (Kd) for TCPP-antibody complex (M)

Antibody Protein	TCPP	TCPP(Fe(III))
Mab 03-1	6.4×10^{-8}	1.5×10^{-7}
L 03-1	2.4×10^{-6}	1.0×10^{-5}
Mab 13-1	1.0×10^{-7} a)	--b)
L 13-1	2.6×10^{-6}	1.4×10^{-5}

a) Kd was determined by ELISA. Other Kd values were by fluorescence quenching.
b) No specific interaction was observed.

Some naturally occurring active antibodies of a camel are composed of heavy chain dimmers and devoid of L chain(*14*). It was also reported that recombinant single immunoglobulin variable domains of H chain secreted from *E. coli* had specific binding activities(*15*). However, specific antigen binding using isolated L chain has not been reported before.

Measurement of peroxidase activity

Since L chains of the Mab 03-1 and Mab 13-1 antibodies exhibit specific interactions with TCPP Fe(III), we are interested in examining whether a peroxidase reaction can be enhanced using the complex of L chain and TCPP Fe(III). Using pyrogallol as a substrate, peroxidase activity was tested for Mab 03-1 and Mab 13-1 L chain complexes with TCPP Fe(III) and compared with those of TCPP Fe(III) alone and Mab 03-1-TCPP Fe(III) (Figure 3a.)

The L chain of Mab 03-1 exhibits slightly higher activity than TCPP Fe(III) but the activity was still lower than that of Mab 03-1. Interestingly, the 13-1 L chain exhibited about 3.5 times higher peroxidase activity than Mab 03-1 (Figure 3a).

These results and the observation that Mab 13-1 did not bind to TCPP Fe(III) suggested that H chain of Mab 13-1 antibody might interfere with the binding of TCPP Fe(III). This resultant complex of 13-1 antibody L chain and TCPP Fe(III) exhibited the highest peroxidase activity among all the tested and previously reported catalytic antibodies. Therefore, we named these chains of antibody fragments with peroxidase activity L-zyme.

Specificity of the peroxidase activity was also examined using other substrates, ABTS, (2,2-azinobis 3-ethylbenz-thiazolin-6-sulfonic acid) (Figure 3b) and catechol. A significant increase in peroxidase activity for ABTS was observed using the complex of 13-1 L chain and TCPP Fe(III) (Figure3b). Yet, no activity was detected when catechol was used as a substrate (data not shown). The

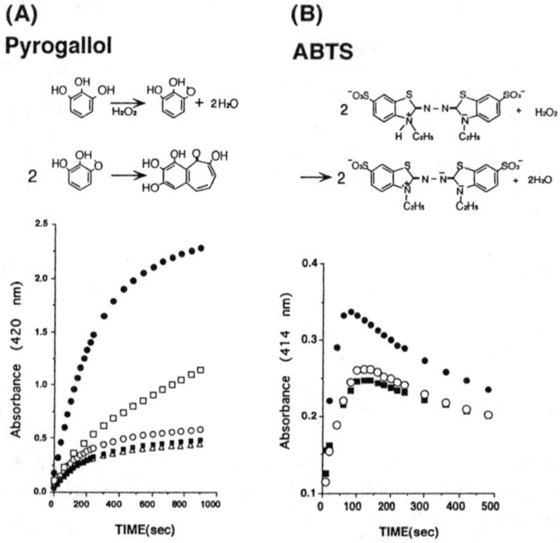

Figure 3. Reaction of pyrogallol (A) and ABTS (B). Peroxidase reaction by complexes of antibody proteins and TCPP Fe(III). Symbols ■ Δ ❑ O and ● indicate TCPP Fe(III) mixed with none, BSA, Mab 03-1, 03-1 L chain, and 13-1 L chain, respectively. The reaction was performed at 37° C as follows. Reaction mixture containing 5.0 x 10-7 M TCPP Fe(III), 8.0 x 10-7 M antibody protein, 1.2 x 10-3 M substrate, and 5.0 x 10-3 hydrogen peroxide dissolved into TAB buffer (90 mM Tris-acetate pH 8.0) was prepared and incubated at 37° C. Absorban ce of wavelength at 420 nm (pyrogallol) or 414 nm (ABTS) was monitored by Shimadzu UV160 UV-visible resording spectrophotometer. The reaction mixtures without antibody protein and with BSA (bovine serum albumin) at the same concentration were prepared as control.

enhancement of peroxidase activity for ABTS might be based on a similar mechanism as explained above for pyrogallol. However, the intensity of enhancement was found to be smaller probably due to the larger size of ABTS (Figure 3b).

Conclusions

We have demonstrated that peroxidase activity could be engineered by the formation of chemically synthesized TCPP Fe(III) and anti-TCPP monoclonal antibody complex. This method provides a new approach to develop a catalytic antibody. Furthermore, it was shown that recombinant L chain could recognize and interact with TCPP and TCPP Fe(III). The complex of 13-1 L chain and TCPP Fe(III) exhibits much higher peroxidase activity than TCPP Fe(III) alone or Mab 03-1 with TCPP Fe(III). We are now examining the molecular mechanism of this new L-zyme through amino acid substitutions and structural studies.

Literature Cited

1. A. Tramontano, K.D. Janda, R.A. Lerner, Science, 234, 1566 (1986).
2. S.J. Pollack, J.W. Jacobs, P.G. Schultz, Science, 234, 1570 (1986).
3. J.W. Jacobs, P.G. Schultz, R. Sugasawara, J. Powell, J. Am. Chem. Soc., 109, 2174 (1987).
4. A.C. Braisted and P.G. Schultz, J. Am. Chem. Soc. 112, 7430, (1990).
5. A.G. Cochran, R. Sugasawara, P.G. Schultz, J. Am. Chem. Soc. 110, 7888 (1988).
6. K.M. Shokat, C.J. Leumann, R. Sugasawara, P.G. Shcultz, Nature. 338, 269 (1989).
7. N. Janjic, a. Tramontano, J. Am. Chem. Soc. 111, 9101 (1989).
8. A.G. Cochran and P.G. Schultz, J. Am. Chem. Soc. 112, 9414 (1990).
9. A. Harada, K. Okamoto, M. Kamachi, Chem. Lett. 953 (1991).
10. R.L. Mullinax et al. Proc. Natl. Acad. Sci. USA. 87, 8095 (1990).
11. F. Sanger, S. Nicklen, A.R. Coulson, Proc. Natl. Acad. Sci. USA. 74, 5463 (1977).
12. F.W. Studier, A.H. rosenberg, J.J. Dunn, J.W. Dubendorff, Methods Enzymol. 185, 60 (1990).
13. G. Scatchard, Ann. N.Y. Acad. Sci. 51, 660 (1949).
14. C. Harmers-Casterman et al., Nature, 363, 446 (1993).
15. E.S. Ward, d. Gussow, A.D. Griffiths, P.T. Jones, G. Winter, Nature 341, 544-546 (1989).

RECEIVED June 13, 1995

INDEXES

Author Index

Affiliation Index

Subject Index

A

Production: Amie Jackowski
Indexing: Deborah H. Steiner
Acquisition: Barbara E. Pralle
Cover design: Cornithia Allen Harris

Printed and bound by Maple Press, York, PA

Highlights from ACS Books

Good Laboratory Practice Standards: Applications for Field and Laboratory Studies
Edited by Willa Y. Garner, Maureen S. Barge, and James P. Ussary
ACS Professional Reference Book; 572 pp; clothbound ISBN 0–8412–2192–8

Silent Spring Revisited
Edited by Gino J. Marco, Robert M. Hollingworth, and William Durham
214 pp; clothbound ISBN 0–8412–0980–4; paperback ISBN 0–8412–0981–2

The Microkinetics of Heterogeneous Catalysis
By James A. Dumesic, Dale F. Rudd, Luis M. Aparicio, James E. Rekoske, and Andrés A. Treviño
ACS Professional Reference Book; 316 pp; clothbound ISBN 0–8412–2214–2

Helping Your Child Learn Science
By Nancy Paulu with Margery Martin; Illustrated by Margaret Scott
58 pp; paperback ISBN 0–8412–2626–1

Handbook of Chemical Property Estimation Methods
By Warren J. Lyman, William F. Reehl, and David H. Rosenblatt
960 pp; clothbound ISBN 0–8412–1761–0

Understanding Chemical Patents: A Guide for the Inventor
By John T. Maynard and Howard M. Peters
184 pp; clothbound ISBN 0–8412–1997–4; paperback ISBN 0–8412–1998–2

Spectroscopy of Polymers
By Jack L. Koenig
ACS Professional Reference Book; 328 pp;
clothbound ISBN 0–8412–1904–4; paperback ISBN 0–8412–1924–9

Harnessing Biotechnology for the 21st Century
Edited by Michael R. Ladisch and Arindam Bose
Conference Proceedings Series; 612 pp;
clothbound ISBN 0–8412–2477–3

From Caveman to Chemist: Circumstances and Achievements
By Hugh W. Salzberg
300 pp; clothbound ISBN 0–8412–1786–6; paperback ISBN 0–8412–1787–4

The Green Flame: Surviving Government Secrecy
By Andrew Dequasie
300 pp; clothbound ISBN 0–8412–1857–9

For further information and a free catalog of ACS books, contact:
American Chemical Society
Product Services Office
1155 16th Street, NW, Washington, DC 20036
Telephone 800–227–5558

Bestsellers from ACS Books

The ACS Style Guide: A Manual for Authors and Editors
Edited by Janet S. Dodd
264 pp; clothbound ISBN 0–8412–0917–0; paperback ISBN 0–8412–0943–X

Understanding Chemical Patents: A Guide for the Inventor
By John T. Maynard and Howard M. Peters
184 pp; clothbound ISBN 0–8412–1997–4; paperback ISBN 0–8412–1998–2

Chemical Activities (student and teacher editions)
By Christie L. Borgford and Lee R. Summerlin
330 pp; spiralbound ISBN 0–8412–1417–4; teacher ed. ISBN 0–8412–1416–6

Chemical Demonstrations: A Sourcebook for Teachers, Volumes 1 and 2, Second Edition
Volume 1 by Lee R. Summerlin and James L. Ealy, Jr.;
Vol. 1, 198 pp; spiralbound ISBN 0–8412–1481–6;
Volume 2 by Lee R. Summerlin, Christie L. Borgford, and Julie B. Ealy
Vol. 2, 234 pp; spiralbound ISBN 0–8412–1535–9

Chemistry and Crime: From Sherlock Holmes to Today's Courtroom
Edited by Samuel M. Gerber
135 pp; clothbound ISBN 0–8412–0784–4; paperback ISBN 0–8412–0785–2

Writing the Laboratory Notebook
By Howard M. Kanare
145 pp; clothbound ISBN 0–8412–0906–5; paperback ISBN 0–8412–0933–2

Developing a Chemical Hygiene Plan
By Jay A. Young, Warren K. Kingsley, and George H. Wahl, Jr.
paperback ISBN 0–8412–1876–5

Introduction to Microwave Sample Preparation: Theory and Practice
Edited by H. M. Kingston and Lois B. Jassie
263 pp; clothbound ISBN 0–8412–1450–6

Principles of Environmental Sampling
Edited by Lawrence H. Keith
ACS Professional Reference Book; 458 pp;
clothbound ISBN 0–8412–1173–6; paperback ISBN 0–8412–1437–9

Biotechnology and Materials Science: Chemistry for the Future
Edited by Mary L. Good (Jacqueline K. Barton, Associate Editor)
135 pp; clothbound ISBN 0–8412–1472–7; paperback ISBN 0–8412–1473–5

For further information and a free catalog of ACS books, contact:
American Chemical Society
Product Services Office
1155 16th Street, NW, Washington, DC 20036
Telephone 800–227–5558